Beyond Imagination

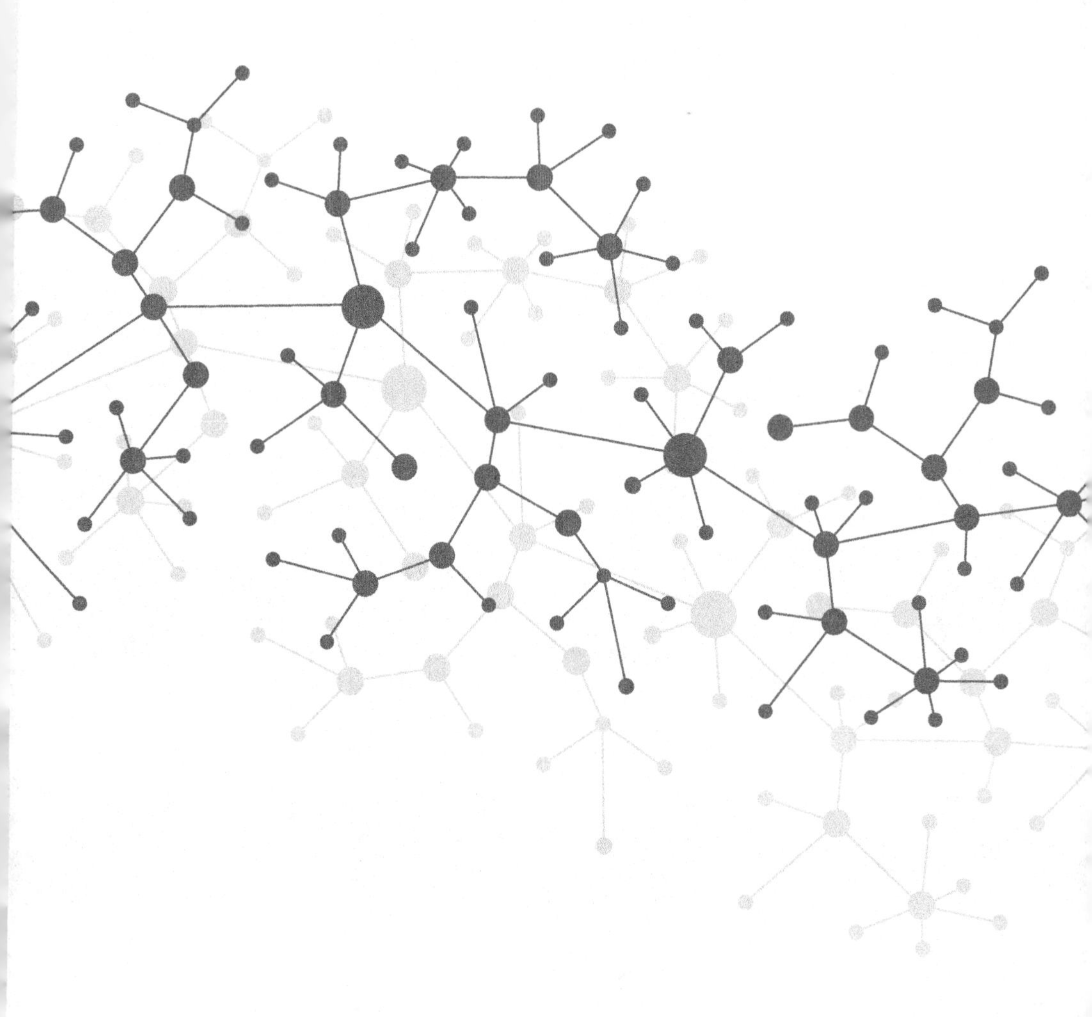

Beyond Imagination

The Ethics and Applications of Nanotechnology and Bio-Economics in South Africa

EDITED BY

Zamanzima Mazibuko

THIS RESEARCH PROJECT WAS SUPPORTED BY:

First published by the Mapungubwe Institute for Strategic Reflection (MISTRA) in 2018

142 Western Service Rd
Woodmead
Johannesburg, 2191

ISBN 978-0-6399238-0-2

Production and design by Jacana Media, 2018
Editor in chief: Joel Netshitenzhe
Text editor: Terry Shakinovsky
Copy editor: Megan Mance
Proofreader: Lara Jacob
Designer: Shawn Paikin

Set in Stempel Garamond 10.5/15pt
Printed and bound by ABC Press, Cape Town
Job no. 003241

When citing this publication, please list the publisher as MISTRA.

Contents

Preface

TECHNOLOGICAL INNOVATIONS ARE CRITICAL for the improvement of the human condition. They also have the potential to improve a country's productivity and competitiveness.

Biotechnology and nanotechnology are the most eminent emerging technologies and are anticipated to be instrumental in addressing global challenges and societal needs. They are reshaping the landscape of science and technology and have become part of national research agendas. Indeed, the science of very small particles – the possibility to reconfigure the arrangement of atoms, the capacity to modify biological cells to improve performance and correct abnormalities – all these present possibilities beyond imagination. But, precisely because of their potency, these technologies also pose risks. It is in the intersection between pursuing benefits and managing risks that exuberance should be tempered by vigilance.

South Africa has embraced the promise of new possibilities that nanotechnology and biotechnology present. This is outlined in relevant national strategies. Led by the Department of Science and Technology, the country seeks to utilise nanotechnology and bio-economics to drive research and development, develop human resources, encourage innovative entrepreneurship, improve demographic inclusivity, improve human health, promote environmental sustainability and ensure food security.

This book, on the application and ethics of biotechnology and bio-economics, seeks to assess the progress that South Africa has made and the lessons it can draw from relevant international experiences.

The authors argue that the application of nanotechnology should be accompanied by an appreciation of nanotoxicity and nanoethics. Similarly, biotechnology should factor in bioethics and the dangers of genetic modification. At the core of this argument is that technology can make maximum social impact only if it brings into play social considerations found in the humanities and social sciences. Transdisciplinarity should be seen as a necessity rather than an afterthought.

While the application of nanotechnology has a long history, going back thousands of years, it is in the context of current technological advancement that the full impact of the technology can be understood. In this context, it is possible to trace the starting material, the invasive routes it follows in a biological setting and ultimately nanowaste. A critical consideration in this regard should be to protect users, researchers and others who interact with nanomaterial.

The book also reflects on the place and role of nanotechnology and bio-economics in a developmental setting. This applies to the orientation of the research, to ensure that it addresses the needs of the majority. It also relates to the role of the private sector and the global partnerships that developing countries can forge with public, private and research institutions. Developing countries, particularly in Africa, should pool their resources and exercise sovereignty so the partnerships they forge address their interests. At the same time, they should develop a corps of skilled personnel with whom the international partners can work. It is against this background that some of the chapters encompass comparative reflections, from which South Africa and the rest of the continent can draw relevant lessons.

The vitality of the chain of research, application and commercialisation is fundamental to maximising the benefit of these technologies. This requires requisite allocation of resources and an appropriate national system of innovation that brings together researchers, government and the private sector.

While South Africa has made appreciable progress in pursuing nanotechnology and bio-economics, it does suffer some deficits. In some instances, the regulations are so rigid as to inhibit partnerships with global entities. In other instances, they are so lax as to allow for unethical conduct. This is besides the issue of availability of resources against the backdrop of competing needs. Further, a critical element in the advancement of these and other technologies is the implementation of the concept of public communication of science, so society can fully appreciate the benefits and the dangers, the career opportunities and the logic behind allocation of resources to these endeavours.

Because of the fledgling nature of these technologies and the depth and breadth of their reach, it is to be expected that a book of this nature can only be exploratory. The Mapungubwe Institute (MISTRA) hopes that it will lay the basis for further reflection on these issues.

MISTRA wishes to thank the authors for their expertise and rigour in interrogating these complex issues in a manner that is accessible. The Department of Science and Technology availed resources for this research initiative, and this is profoundly appreciated.

– Joel Netshitenzhe
Executive Director

Acknowledgements

THE MAPUNGUBWE INSTITUTE for Strategic Reflection would like to express its sincere gratitude to the project coordinators, Radhika Perrot and Zamanzima Mazibuko, and to Velaphi Msimang who supervised the project. We also express our thanks to the researchers who contributed to the body of work.

Thank you to the MISTRA staff who contributed to the successful outcome of this project: the project management directorate led by Xolelwa Kashe-Katiya supported by Dzunisani Mathonsi; Wandile Ngcaweni and Temoso Mashile for assisting with assembling the manuscript; Lorraine Pillay for fundraising and financial management activities with support from Magati Nindi-Galenge; Terry Shakinovsky for editing the book, and Barry Gilder for his continuous efforts to ensure that the publication meets the highest standards. Deepest gratitude goes to Joel Netshitenzhe for his thorough reading of the manuscript.

MISTRA extends its appreciation to Jacana Media, which was responsible for the design, layout and production of the book.

MISTRA also expresses its heartfelt thanks to the Department of

Science and Technology (DST) without whose generous donation this publication would not have been possible.

MISTRA FUNDERS AND DONORS

Though the following donors and funders have not been directly involved with this research, MISTRA would like to thank them for their support. They include:

- ABSA
- Airports Company of South Africa Limited (ACSA)
- Albertinah Kekana
- Anglo American
- Anglo Coal
- Anglo Gold Ashanti
- Anglo Platinum
- Aspen Pharmacare
- Batho Batho Trust
- Belelani Group
- Brimstone
- Chancellor House
- Discovery
- First Rand Foundation
- Friedrich-Ebert-Stiftung (FES)
- Goldman Sachs
- Harith General Partners
- Jackie Mphafudi
- Kumba Iron Ore
- Mitochondria
- National Institute for the Humanities and Social Sciences (NIHSS)
- National Lotteries Commission (NLC)
- Oppenheimer Memorial Trust (OMT)
- Open Society Initiative South Africa (OSISA)
- Phembani Group
- Power Lumens Africa
- Royal Bafokeng Holdings

- South African Breweries
- Safika
- Shell South Africa
- Simeka
- Standard Bank
- Vhonani Mufamadi
- Yellowwoods

List of contributors

Radhika Perrot is a freelance researcher with experience in technology policy and research analysis of climate change, renewable energy technologies and innovation policies. She has researched and worked on various issues and topics around renewable energy, namely, feed-in policies and socio-technological factors behind solar PV, wind and hydrogen-fuel cells innovation, and understanding market competition and firm strategies in global markets. Her current research focus is on comparing the transition of energy systems to low-carbon technologies in the BRICS countries looking at the barriers and drivers of innovation of these technologies. Her PhD thesis has analysed strategic innovation activities of global firms over the technology life cycle of renewable energy technologies. She has contributed to various journals with policy, academic and market analysis.

Zamanzima Mazibuko is a senior researcher in the Knowledge Economy and Scientific Advancement Faculty at MISTRA. She is a Wits university alumni and holds a BSc degree in Biochemistry and Cell Biology and a BSc honours in Pharmacology. She obtained her MSc

(Med) in Pharmaceutics cum laude and has published on nano-enabled drug delivery technologies in Amyotrophic Lateral Sclerosis. She has a patent filed with Wits Enterprise for a nano-enabled drug delivery system designed and formulated for her master's degree. Zamanzima's current research interests are on the low-carbon economy; beneficiation of strategic minerals in South Africa, particularly platinum group metals; nanomedicine; and epidemics and health systems in Africa.

Puleng Matatiele is currently a senior medical scientist in the Analytical Services section of the National Institute for Occupational Health (NIOH), Johannesburg. Previous to that, she was a medical scientist at the Toxicology section of the NIOH. Puleng Matatiele has published over 10 peer-reviewed research papers and review articles, on a variety of topics including breast cancer, nanoparticles and green technologies.

Her broad research interests are at the interface between toxicology and analytical chemistry with a particular focus on human health and biomarkers of chemical exposure. Otherwise, Puleng is a bookworm; she devours books and articles on health, diet, fitness, entrepreneurship, personal finance, personal development and prayer.

Natasha Sanabria is currently at the National Institute of Occupational Health (NIOH/NHLS) and Research Unit in Bioinformatics, Rhodes. She investigates biochemical and genetic analyses of disease-related states, such as gene expression during stressed conditions, 'Self/non-self' recognition events, innate immunity, cellular signal perception/transduction, as well as assessment of nanomaterial toxicity. Dr Sanabria worked part-time while completing her Biochemistry MSc degree, which was awarded cum laude in 2003 (RAU). She received the NRF Prestigious award, discovered a new gene (GenBank accession number GU196248) and obtained her PhD in 2009 (UJ). Thereafter, Dr Sanabria received the NRF Innovation fellowship to complete postdoctoral studies in 2011, with training at Cold Spring Harbour Laboratory (New York, USA). Dr Sanabria worked as the Biochemistry honours-course coordinator in 2012, before transferring to the Metacatalysis/'Metals in medicine' units at UJ to work as the QC manager. Dr Sanabria moved to the NIOH/NHLS in 2013 as the

Toxico-genomics Lab Manager and held an NHLS Research Trust Development Grant (2015–2017). She also served as the Coordinator for the WHO Exposure Assessment to NPs Workshop (2013), NanOEH conference organising committee member (2015), higher-degrees co-supervisor, as well as mentor for DST/NRF industry interns (2015/2016). Currently, Dr Sanabria is pursuing an MSc in Bioinformatics and Computational Molecular Biology.

Melissa Vetten completed her MSc in Biochemistry from the University of Johannesburg in 2009. She has been employed as a scientist at the National Institute for Occupational Health, South Africa, since 2008 where she has worked on numerous research projects including an investigation on the mechanisms involved in increased susceptibility to Mycobacterium tuberculosis in silica-exposed individuals, and a study investigating the in vitro toxicity of dust collected from mines in South Africa. Since 2010, she has been involved in nanotoxicology research. She completed the South African contribution to the OECD Working Party on Manufactured Nanomaterials (WPMN) Colony Forming Efficiency assay interlaboratory study. She is currently completing her PhD at the University of the Witwatersrand, investigating the in vitro toxicity of manufactured gold nanoparticles in a bronchial epithelial cell line.

Mary Gulumian is the head of the Toxicology and Biochemistry section, National Institute for Occupational Health (NIOH). She also holds an honorary professorial post in the Haematology and Molecular Medicine Department, University of the Witwatersrand, where she presents courses on Health Risk Assessment (nanotechnologies) and supervises postgraduate students. She is the founder member and the past president of the Society for Free Radical Research of South Africa (SFRR-SA) and the founder member and president of the Toxicology Society of South Africa (TOXSA). She served in the executive committee (2013–2016) of the International Union of Toxicology (IUTOX) as vice president. She was a member of the final review board of WHO Concise International Chemical Assessment Documents (CICAD) publications on number of toxic compounds. She currently represents South Africa in the OECD Working Party on Manufactured Nanomaterials as well as serves on Expert Technical Committee for WG3 of ISO/TC 229 on nanotechnologies. Her research interests include hazard identification and communication as well as elucidation of mechanisms of toxicity of micro and nano particles. She has authored and co-authored numerous scientific publications on this topic and given presentations at local and overseas conferences. She has also provided expert consultation to industry and government departments on the toxicity of chemicals in working and ambient environments. Professor Gulumian is also a member of the editorial board of the journal, *Human and Experimental Toxicology* (HET) and also of the journal, *Particle and Fibre Toxicology* (PFT).

Hailemichael Teshome Demissie is an associate professor of Law at the School of Law of the University of Gondar in Ethiopia. He was a senior research fellow of the Nairobi-based intergovernmental think tank, the African Centre for Technology Studies (ACTS) following his research fellowship at Tshwane University of Technology (TUT) in Pretoria, South Africa. He holds a PhD from King's College London and an LLM with distinction from the University of Warwick. Dr Demissie has written extensively on issues emerging on the interface between law and technology, on technology policy, ethics and regulation. His latest book, co-edited with Dr Cosmas Ochieng, Dr Guillermo Foladori and Dr Desalegn Mengesha, is published by ACTS Press and the University of Gondar Press. The book, *Harnessing Nanotechnology for Sustainable Development in Africa*, has already received several positive reviews. Dr Demissie's other recent works include the successful organisation of two global conferences on 3D printing technology at Kenyatta University in Nairobi, Kenya and at the University of Gondar in Gondar, Ethiopia. He is currently leading a research project on '3D printing for accelerated economic development in Africa'. His contributions on the same theme published on the SciDev website have been widely cited across the globe.

Thomas S Woodson is an assistant professor at Stony Brook University in the Department of Technology and Society. He specialises in science and technology policy and international development. For his current projects, Thomas is investigating the effects of technology on inequality throughout the world and the causes/consequences of inclusive innovation. Dr Woodson received his bachelor's degree in Electrical Engineering from Princeton University and his PhD in Public Policy with a specialisation in science and technology policy from the Georgia Institute of Technology (Georgia Tech). While at Georgia Tech, he was a part of the Technology Policy Assessment Center and the Center for Nanotechnology in Society at Arizona State University. At these centres, he investigated the effects of nanotechnology on inequality and poverty throughout the world. Dr Woodson received numerous awards while at Georgia Tech including the Georgia Tech Presidential Fellowship and the NSF Graduate Student Fellowship.

Shih-Hsin Chen is a faculty member at the Institute of Management of Technology (AACSB accredited) at National Chiao Tung University. She has a PhD degree in Science and Technology Studies from the University of Nottingham, UK. Her research interests lie in social network analysis, system of innovation, technology management, technology assessment, and technology policy regarding the development of emerging technologies.

Swapan Kumar Patra has a PhD in Science Policy from the Centre for Studies in Science Policy, School of Social Sciences, Jawaharlal Nehru University, New Delhi, India. Dr Patra is presently working as a postdoctoral research fellow at Tshwane University of Technology, Pretoria, South Africa. He is also associate editor of the *African Journal of Science Technology, Innovation and Development* (AJSTID) and *Tshwane University of Technology – Journal of Creativity, Innovation and Social Entrepreneurship* (TUT-JCISE). His current research interests are globalisation of research & development with particular reference to India and China, multinational enterprises, open innovation, sectoral system of innovation, system of innovation in India, China, South Africa and other developing countries. Dr Patra extensively uses network analysis, scientometrics and bibliometrics, science and technology indicators, and statistical tools in his research. He has published a number of journal articles and book chapters.

Mammo Muchie did his undergraduate degree at Columbia University, and his MPhil and DPhil at the University of Sussex, UK. He is currently a DST/NRF research professor of Innovation Studies, at Tshwane University of Technology, Pretoria, South Africa. He is also a NRF-rated research professor. He is a fellow of the South African Academy of Sciences, the African Academy of Sciences and the African Scientific Institute. He is currently an adjunct professor at the Adama Science and Technology University, Addis Ababa University, University of Gondar and Arsi University, Ethiopia. He has been a senior research associate at the SLPMTD programme and now has also become a senior research associate at the TMCD Centre of Oxford University collaborating with researchers on creativity and innovation in low-income economies. He has held various positions, including the director of the Research Programme on Civil Society and African Integration at the then University of KwaZulu-Natal, professor and founding director of the Research Centre on Innovation and International Political Economy and The Comparative Research Centre on Integration, principal lecturer at Middlesex University, professor at Aalborg University, part-time lecturer at Cambridge University, and many more. He is currently the chairman of the advisory board of the African Talent hub of the Community Interest Company to raise funds for making Africa the talent, innovation, entrepreneurship, creativity and knowledge hub of the world.

Shashank S Tiwari is a scholar of Science and Technology Studies (STS). He did his MPhil in Science Policy Studies from Jawaharlal Nehru University, India and PhD in STS from the University of Nottingham, United Kingdom. He specialises in the social studies of life sciences, biomedicine and biomedical technology. Dr Tiwari is currently based in the United States of America and works as a science policy analyst.

Abbreviations

ADME	absorption, distribution, metabolism and excretion
AIIMS	All India Institute of Medical Sciences
AMTS	Advanced Manufacturing Technology Strategy
ATCA	Alien Tort Claims Act
AU-NEPAD	African Union-New Partnership for Africa's Development
BAC	Biotechnology Advisory Committee
BNPDA	Biotech and New Pharmaceutical Development Act
BRICs	Biotechnology Regional Innovation Centres
BRICS	Brazil, Russian, India, China and South Africa
BTC	Biotechnology Committee
CCMB	Centre for Cellular and Molecular Biology
CDSCO	Central Drugs Standard Control Organization
CeSTII	Centre for Science, Technology and Innovation Indicators
CIA	Central Intelligence Agency
CMC	Chemistry, Manufacturing and Controls
CNF	carbon nanofiber

CNT	carbon nanotubes
CRO	contract research organisation
CSIR	Council for Scientific and Industrial Research
DALY	disability adjusted life years
DBT	Department of Biotechnology
DCB	Development Centre for Biotechnology
DCGI	Drug Controller General of India
DWCNT	double-walled carbon nanotube
DNA	deoxyribonucleic acid
DPN	The dip-pen nanolithography
DSIR	Department of Scientific and Industrial Research
DST	Department of Science and Technology
EHS	Environmental Health and Safety
ELSA	ethical, legal and social aspects
ENM	engineered nanomaterials
ESR	electron spin resonance
EU	European Union
FAP	Focus on Africa Programme
FDI	Foreign Direct Investments
FICCI	Federation of Indian Chambers of Commerce and Industry
GD	gestational days
GDP	Gross Domestic Product
GEAR	Growth, Employment and Redistribution
GERD	Gross domestic expenditure on R&D
GI	gastro-intestinal
GM	Genetically modified
GMO	genetically modified organism
GoI	Government of India
GPT	general purpose technologies
hESC	human embryonic stem cell
HIV	human immunodeficiency virus
HIV/AIDS	human immunodeficiency virus/acquired immune deficiency syndrome
HSRC	Human Sciences Research Council
IAVI	International Aids Vaccine Initiative

ICAR	Indian Council of Agriculture Research
ICMR	Indian Council of Medical Research
ICT	information and communication technology
IHME	Health Metrics and Evaluation
IPAP	Industrial Policy Action Plan
IPC	International Patent Classification
iPS	Induced Pluripotent stem cells
IR&C	International Relations and Cooperation
IS	innovation systems
ISO	International Standards Organization
ISSL	International Stem Cell Services Ltd
IT	information technology
IWGN	Interagency Working Group on Nanotechnology
JNCASR	Jawaharlal Nehru Centre for Advance Scientific Research
KEI	Knowledge Economy Index
LCA	Life Cycle Assessment
LGP	Life Guard Pharmaceuticals
LVPEI	L.V. Prasad Eye Institute
MCC	Medicine Council Control
MDG	Millennium Development Goals
MNE	multinational enterprises
MNT	molecular nanotechnology
MRC	Medical Research Council
MRSCA	Medicines and Related Substances Control Act
MSC	mesenchymal stem cells
MSIDC	Microsoft India Development Centre
MWCNT	multi-walled carbon nanotubes
NACI	National Advisory Council on Innovation
NBB	National Biotechnology Board
NBIC	Nano, Bio, Info and Cogno
NBS	National Biotechnology Strategy
NCRM	Nichi-In Centre for Regenerative Medicine
NEPAD	New Partnership for Africa's Development
NGP	New Growth Path
NDF	National Development Fund

NDP	National Development Plan
NE^{3}LS	nano-ethical, environmental, economic and legal and social issues
NELSI	nano-ethical, environmental, economic and legal and social issues
NHA	National Health Act
NIMS	Nizam's Institutes of Medical Sciences
NIS	National Innovation System
nm	nanometer
NNEP	National Nanotechnology Equipment Programme
NNI	National Nanotechnology Initiative
NNS	National Nanotechnology Strategy
NP	nanoparticles
NPEP	Nanotechnology Public Engagement Programme
NRF	National Research Foundation
NRP	National Research Plan
NSI	National System of Innovation
NTC	national technological capabilities
OECD	Organisation for Economic Co-operation and Development
PATH	Programme for Appropriate Technology in Health
PDP	product development partnerships
PGI	Post Graduate Institute of Medical Sciences
PND	postnatal days
ppm	parts per million
PPP	public–private partnership
PRD	poverty related diseases
R&D	research and development
RDP	Reconstruction and Development Programme
RI	relative impact
RLS	Reliance Life Science Pvt. Ltd.
RoC	Republic of China
ROS	reactive oxygen species
S&T	science and technology
SADC	Southern African Development Community
SAMRC	South African Medical Research Council

SCI	Stem Cell Research Initiative
SEIN	social ethical implications/interactions of nanotechnology
SGPGI	Sanjay Gandhi Post Graduate Institute of Medical Sciences
SIDA	Swedish International Development Cooperation
STEM	science, technology, engineering and mathematics
STI	science, technology and innovation
STM	scanning tunnelling microscope
SUI	statute for upgrading industries
SVIMS	Sri Venkateswara Institute of Medical Sciences
SWCNT	single-walled carbon nanotubes
TB	tuberculosis
TEM	transmission electron microscopy
TIA	Technology Innovation Agency
TIP	Technology and Innovation Policy
TYIP	Ten-Year Innovation Plan
UGC	University Grants Commission
UK	United Kingdom
UN	United Nations
UNESCO	United Nations Educational, Scientific and Cultural Organization
USPTO	United States Patent Trademark Office
WISH	Wistar Institute, Susan Hayflick
WoS	Web of Science
WTO	World Trade Organization
XRPD	X-ray powder diffraction

INTRODUCTION

The dynamics of new and emerging technologies in developing countries and the new role of the state

An Introduction

RADHIKA PERROT

NEW AND EMERGING TECHNOLOGIES such as biotechnology and nanotechnology have been heralded as 'disruptive' for they have opened new vistas in multi-disciplinary fields of research and development (R&D), and have been applied in the invention and production of new products and processes across sectors. According to Christensen (1997) in his best-selling book, *The Innovator's Dilemma*, disruptive technologies are scientific discoveries that change the usual product/technology paradigms and provide a basis for a new and more competitive one. Products or innovations based on disruptive technologies create new markets and value networks that eventually disrupt established markets and leading firms, products and alliances. A wave of new, 'disruptive' technological change – such as the internet and computer technologies, biotech and nanotechnology – has had and will continue to have critical consequences for all countries.

The development and diffusion of disruptive technologies are considered vital to any country's economic growth and development because these technologies have beneficial applications in a wide range of sectors. These include healthcare/medicine, electronics, textiles, agriculture, construction, water treatment, food processing and cosmetics – promising to improve existing products and make them cheaper and more effective. Moreover, these disruptive technologies – sometimes called general purpose technologies (GPTs) – have the potential to address many societal challenges that pertain to developing countries. There are three core qualities that characterise GPTs (Mazzucato, 2011: 54):

> *They are pervasive in that they spread to and are applied in products and processes in many sectors.*
> *The performance gets better over time and the cost to the users keeps getting lower.*
> *They easily form the core technological basis in the invention and production of other new products or processes.*

In recent years, technological advancement and development have become increasingly 'systemic' in nature (Teece, 1996). Moreover, the new and emerging technologies are knowledge-intensive in nature, requiring knowledge that can be applied across industries and across borders for the co-development of processes, products and engineering designs and/or acquiring them from other countries or organisations. The 'combinatorial' nature of the technologies (Mytelka, 2004) is causing a 'reorganizing of entire industries' (Fagerberg, 2010: 7) and the diversity of application areas for a given technology has increased in scope.

SOCIOTECHNICAL IMAGINARIES[1]

The pursuit of technosciences such as medicine, agricultural biotechnologies, information and communication technologies (ICTs),

1 Sociotechnical imaginaries are cultural representations and tools to anticipate the future and are often expressed as visions of science in society and visions of science and technology governance.

nanotechnologies and energy technologies is an integral part of the peculiarity of a country's socio-political and institutional context. Research shows that government policies aimed at introducing and promoting new disruptive technologies have been a product of imagination rather than systemic and strategic decision-making, which takes account of the existing sociotechnical and techno-economic environment (Jasanoff & Kim, 2009; Goven & Pavone, 2015). Sociotechnical imaginaries are those 'collectively imagined forms of social life and social order reflected in the design and fulfilment of nation-specific scientific and/or technological projects' (Jasanoff, 2015: 120).

Sociologists such as the social constructivists and the actor-network theorists affirm that technological choices are never purely economic and technological. This is because political and cultural factors also act to 'pattern the design and implementation of technology' (Williams & Edge, 1996). Sociotechnical imaginaries constitute political–institutional change that shapes the parameters of possible future action (Goven & Pavone, 2015).

Though collectively held, it is possible for sociotechnical imaginaries to originate in the vision of a single individual, a group of scientists, or politicians or the specific needs of communities. In South Africa, sociotechnical imaginaries of new and emerging technologies are held by the Department of Science and Technology (DST). Its mandate since its institution has been to *socialise* science, innovation and technologies – those new innovations and technologies that hold the potential to address the country's societal challenges such as inadequate access to healthcare, unemployment and income inequality. Mazzucato (2011) argues that what is critical is not funding so-called blue-sky research, but rather creating *visions* around important new technologies.

The visionary nanotech revolution of the US government is an example of a sociotechnical imagination, the vision and efforts of which were led by a small group of scientists and engineers at the National Science Foundation and the Clinton White House in the late 1990s (Motoyama, Appelbaum & Parker, 2011).

SCIENCE & TECHNOLOGY: PRO-POOR STRATEGIES?

There are a number of developing countries that are pursuing the research and development of biotechnology and nanotechnology strategies, namely South Africa, India, China, Mexico, Turkey, Brazil and Argentina. The biotechnology and nanotechnology strategies of most of these developing countries have pro-poor elements in their design – arguing that these technologies will respond to the socio-economic challenges of their countries, namely alleviating poverty, reducing unemployment and improving health.

However, emerging technologies do not usually improve the lives of people at the bottom of the global income distribution, but instead have been known to accumulate wealth at the top (Cozzens, 2012). High-tech technologies such as nanotechnology might not initially appear critical for developing countries. For example, the citations for and commercialisation of nanotechnology products are heavily concentrated in affluent countries (Cozzens, 2012). In addition, developing countries can be shown to have managed without such technologies. Invernizzi and Foladori (2005) have noted the ability of China and Vietnam to significantly reduce malaria in the last century without the use of emerging technologies.

It is important for developing countries' specific science and technology (S&T) strategies to consider who will benefit from the product, and even more important to get to know the needs of the people, and assess emerging technologies in relation to those needs (Cozzens, 2012).

Globally, countries are seeking innovation-led growth that is more 'inclusive' and 'sustainable' than in the past (Mazzucato, 2013a). But policies that promote growth at the expense of increasing income inequality are not pro-poor. Concentrating on such policies runs the risk of ignoring overall economic welfare and even the fortunes of the nearly-poor (Page, 2005). Further, according to Page (2005), policies to spur growth can result in increases in income inequality but nonetheless remain pro-poor, as long as they raise the overall incomes of the poor.

State policies aimed at fostering S&T development should clearly continue to emphasise basic research, but not to the exclusion of supporting payoffs to the innovative producers or market entrepreneurs (Appelbaum, 2016). Furthermore, state policies should ensure that these emerging technologies benefit the poor, by directly serving their needs and/or improving their income through promoting pro-poor market strategies or pro-poor growth.

Past S&T policies of developing countries were dependent on technologies and innovations from developed countries, but failed to address societal challenges or to benefit the majority of the population, mostly the poor. However, recently there has been recognition of the fact that for adopted technologies to have required economic and social impact, a certain level of 'absorptive capacity'[2] (Cohen & Levinthal, 1990) was required. The adoption and acquisition of new technologies also required behavioural and systemic changes at the level of the 'roll out' of these technologies by the government.

The interdisciplinary nature of nanotechnology also poses problems for researchers and institutions used to traditional disciplines with well-defined boundaries (Tegart, 2006). Nanotechnology will require the changing of traditional mindsets, which is a major challenge, and a particular need is to develop nanotechnology experts with interdisciplinary skills in research and policy. According to Marchant and Wallach (2015), no single entity is capable of fully governing any of these multi-faceted and rapidly emerging fields; a diverse set of governance actors, programmes, instruments and influences apply to each form of technology introduced.

THE ENTREPRENEURIAL STATE

Living in today's era 'requires a new justification of government intervention that goes beyond the usual one of "fixing market failures"' and should extend to the shaping and creation of markets (Mazzucato, 2013a: 3). In her provocative book, *The Entrepreneurial State*, Mazzucato argued for a far more proactive role for government

---

2 Absorptive capacity is a firm's ability to identify, assimilate, transform and apply valuable external knowledge.

than that of simply 'fixing markets', providing 'the conditions for innovation' and investing in skills and in getting a strong science base to flourish. She urges governments to invest where the private sector will not, usually in the most uncertain and risky areas of research and innovation. The concept of *entrepreneurial risk-taking* by the state is introduced, and her book provides many examples from the ICT, pharmaceutical and biotech industries where it has been the state and not the private sector, that created the much-needed economic dynamism. In fact, Mazzucato (2011: 22) observed that:

> *Risky research is funded by the publicly funded labs (the National Institutes of Health or the MRC) while private pharma focuses on less innovative 'me too drugs' and private venture capitalists enter only once the real risk has been absorbed by the state. And yet make all the money. In industries with such long time horizons and complex technologies, it is argued that return-hungry venture capital can in fact sometimes be more damaging than helpful to the ability of the sector to produce valuable new products.*

The role of the government should go beyond basic blue-sky research, and the current role of creating knowledge and a skills base through national labs and universities: Government's role should include resource mobilisation and creating the conditions for widespread market commercialisation (Mazzucato, 2011; Cozzens, 2012). In the latter case, the government can allow knowledge and innovation to diffuse across sectors and throughout the economy either through existing networks or by facilitating new ones (Mazzucato, 2011). In addition, governments should ensure that as much knowledge as possible is accessible to market entrepreneurs, social and otherwise (Cozzens, 2012).

INNOVATION-LED GROWTH AND COLLABORATIONS

Innovation-led growth

In seeking innovation-led growth, it is fundamental to understand the important roles that both the public and private sectors can play. It is critical to understand and rethink *what it is that the public and the private sectors can bring to the ecology* (Mazzucato, 2011). The idea of the entrepreneurial state suggests that one of the central missing links between growth and inequality lies in a wider identification and understanding of the agents that contribute to the risk-taking[3] that is required for that growth to occur.

The innovation systems (IS) framework provides a lens for the identification and understanding of the role of the various agents that contribute to innovation-led growth. This framework also offers a systemic approach, with insights into the innovative and economic performances of countries. According to the OECD (1997), the approach is built on the basis that innovation and technology development are the result of a complex set of relationships among actors in the system, which includes enterprises, universities and government research institutes, and involves the flows of technology, knowledge and information among people, enterprises and institutions that are key to the innovative process.

The IS concept was developed in the late 1980s by scholars in Europe (Freeman & Lundvall, 1988; Lundvall, 1992). They argued that in order to understand innovation and learning, it would be important to understand how linkages are formed and interactions between organisations at the national level take place (Lema et al, 2014). And in fact, this 'systemic' approach is expected to provide developing countries with useful theoretical insights into understanding innovation-led

3 See Mazzucato (2011) for an in-depth understanding of the imbalance between the different risk-takers of innovation and the rewards and incentives they receive in return for the risks taken. Further, Mazzucato reiterates that the real risk-taker of basic research and innovation has been governments throughout the various technological revolutions and not the private sector or private venture capitalists as is primarily assumed.

growth and formulating technology-specific innovation policies (Kraemer-Mbula & Wamae, 2010).

However, according to Mazzucato (2011), having a national innovation system (NIS) that is rich in horizontal and vertical networks and linkages is not sufficient to lead a country to an inclusive and innovation-led growth. Rather, the state must play a leading role in the process of development from envisioning strategies for science and technology research to enabling the diffusion of these technologies and innovations in the market.

Innovation via collaborations

Research and development of disruptive technologies such as nanotechnology and biotech are 'combinatorial' or 'systemic' or 'synergistic' in nature and therefore do not take place in isolation but rely on collaboration, much of which takes place across borders and between organisations (Mytelka, 2004; Van Horn & Fichtner, 2008; Aydogan-Duda, 2012; Appelbaum, 2016).

For developing countries, technological change occurs primarily through learning based on the acquisition, diffusion and upgrading of technologies that already exist in more advanced countries – and not by pushing (or even attempting to push) the global knowledge frontier further (Bell & Pavitt, 1995).

Learning and continuous innovation has been the key activity for building technological capability and achieving technological competitiveness, and this has become even more important in developing economies (Adelowo et al, 2015). Learning within the innovation system occurs via interactions or collaborations and partnerships that assist in the flow of knowledge and information between firms and organisations. A core idea of the innovation system is that innovation or novelty depends on interaction among actors with related but different knowledge, and without this variety or difference there is a risk of myopia and of missing out on spotting new opportunities (Lema et al, 2014).

Learning and innovation in the global nanotechnology industry, according to Appelbaum (2016), is comprised of global science and engineering networking and the opportunities arising out of such interactions extend beyond national borders. Such international partnerships are of particular benefit to developing countries because they contribute to technology, knowledge transfer and scientific advancement.

Appelbaum (2016) analysed advancement in nanotechnology globally, and also interviewed several national officials within the context of IS of various developing countries (China, Mexico, Brazil and Argentina), and compared these countries to the USA. The study concluded the following, all of which supports the main premise made by Mazzucato (2011):

> *As the case of China has shown, public investment is not sufficient for a successful innovation system; there are cultural and institutional barriers that need to be overcome in order to translate basic research into commercial success.*

In Mazzucato (2011), Bill Gates is quoted as saying that

> *the key element to get a breakthrough is more basic research and that requires the government to take the lead. Only when that research is pointing towards a product then we can expect the private sector to kick in.*

There are a variety of reasons why commercialisation is often unsuccessful in developing countries: lack of a supportive institutional and legal structure, lack of vision, and cultural idiosyncrasies alike (Aydogan-Duda, 2012). The lessons of Latin America – particularly Mexico, Brazil and Argentina – show that in the absence of strong governmental programmes in nanotechnology, sustained innovative breakthroughs are unlikely even where basic research has some strengths. Such countries are likely to be 'takers' of economically advanced countries' S&T efforts, producing outputs that are at the low end of the value chain (such as nanomaterials and nano-intermediates). Coordinated government programmes would increase the likelihood of successfully moving up the value chain to achieve more innovative (and competitive) breakthroughs.

Modern research does not take place in a vacuum, but relies on collaboration, much of which takes place across borders. The cases of Brazil and Argentina illustrate the need for links to industry: these countries have well-developed scientific and academic sectors, but weak ties with industry have impeded the commercialisation of research, and exploitation of local knowledge. There is a lack of nanotechnology-specific risk capital and equipment sectors, which has made scientific knowledge exploitation more difficult in some of these countries.

Mazzucato (2011) recommends that where breakthroughs have occurred as a result of targeted state interventions for specific companies or technologies, the state should reap some of the financial rewards over time, by retaining ownership of a small proportion of the intellectual property created.

ETHICS OF TECHNOLOGY: TO PURSUE OR NOT TO?

Many emerging technologies still in the early stages of development, such as nanotechnology and biotechnology, including embryonic stem cells, regenerative medicine and artificial intelligence, have given rise to a complex mix of benefits and uncertainties. These technologies have raised public concerns about the potential risks of applications to human health.

In the last decade, nanotechnology entered the policy arena of many countries as a technology that is simultaneously promising and threatening (Beumer & Bhattacharya, 2013). However, the experience of agricultural biotechnology with near-fatal red flags raised quite late into the development process clearly stands as an example not to be followed (Cozzens, 2012).

However, the medical uses of biotechnology generally raise different concerns from those that arise from agricultural applications (Gaskell & Bauer, 2001), even though the scientific basis for these technologies could be similar.

According to Sandler (2009), the functions of government intersect with the ethical and value dimensions of new and emerging technologies in several ways:

S&T policy and funding involve decisions about which endpoints should receive priority and how resources should be allocated in pursuit of those ends. In each case, the policy is intended to accomplish certain goals and its justification therefore depends on these goals being valued more highly than their alternatives. Decisions about priorities are based on value judgements.

Regulation of S&T is intended to accomplish that which is considered to be worthwhile, and justifies any associated costs. Regulation has power, control, oversight and responsibility dimensions, and like policy, regulation has *ineliminable* value components. Regulation includes domains as diverse as facilities permitting (e.g., nuclear power plants and waste-transfer stations), setting research limits (e.g., human subject research and reproductive cloning), risk management (e.g., workplace safety and environmental pollution) and technology use (e.g., privacy protection and non-therapeutic use of human growth hormone).

Government can support research on, raise awareness of and promote responsiveness to social and ethical issues associated with technology (as many believe to be the case with the Human Genome Project). The government can also obscure social and ethical issues associated with technology (as many believe to be the case with genetically modified crops).

Informing and incorporating public perceptions of new and emerging technologies is critical as public perception influences national-level policy, and decisions on funding and advancing such technologies. An example is the case of genetically modified (GM) crops in Europe: the market size and scope for this technology significantly dropped as a result of public opinion on the ills of consuming GM food and crops.

Gastrow et al (2016) surveyed public perception of biotechnology in South Africa. They discovered that between 2004 and 2015 there was a substantial increase in public awareness of biotechnology from 21% of the population to 53%, with a major increase in attitudes that favoured the purchase of GM food. However, although public awareness that GM foods form a part of their diet more than tripled from 13% to 48%, for that same period, the public in general lacked information and awareness of GM food. This seeming contradiction came about because those surveyed were not part of the majority of the population consuming GM food.

According to Cozzens (2012), the global governance processes emerging to deal with the Environmental Health and Safety (EHS)[4] risks of nanotechnology are tilted towards the voices and needs of the global North. Moreover, most capacity is devoted to building and regulating nanotechnology. It takes technical capacity to regulate a technology: capacity to participate in international regulatory discussions; capacity to educate local officials about appropriate regulations; and capacity to implement and enforce them.

Further, according to Cozzens (2012: 129):

> *In the North then EHS researchers have been running to catch up with industry, and regulators have been running to catch up with the research. In the global South, the research part of that scenario is just emerging, and regulation has barely started the race.*

OUTLINE OF THE CHAPTERS IN THIS BOOK

Chapter 1 examines the origins of nanotechnology and traces the earliest application of nano-based technology from the cosmetics (such as hair-dye and eye makeup) used by ancient Egyptians and Mayans to the modern-day applications of carbon nanotube computers and the improvement of the efficiency and quality of existing materials (such as textiles, cosmetics, processors and chemicals). The chapter discusses the ethics (risks and benefits) of nanotechnology, and gives an overview of the uses of technology globally, including technologies that have

---

4 Very little is known about the interaction between human-made nanostructured materials (nanotechnology) and living organisms, which led to global debate about the environmental, health and safety issues surrounding nanotechnologies.

potentially far-reaching socio-cultural and economic implications.

Chapter 2 proposes a number of risk-management strategies to address the current lack of risk-assessment tools in testing the toxicity of nanomaterials. The rate of nanomaterial research, development and production has far exceeded the rate of testing to evaluate their toxicity.

Chapter 3 explores ethical issues around the use of nanotechnology-based products, and discusses *nano-divide* as an ethical issue that might arise due to research and regulatory inaction and incapacity on the part of African nations. There is a growing gap between nanotechnology research in countries of the global North and African countries in the global South, with the latter often having only limited means to take advantage of rapid technological development and application.

Chapter 4 analyses the role of private–public partnerships (PPPs) for nano-medicine development in South Africa especially in relation to global diseases of poverty. But the chapter finds that the tight regulations of the South African government are slowing down the ability of such partnerships to operate effectively. These tight regulations will greatly hinder innovation in the country unless the government re-examines its regulations and policies to encourage research and innovation of PPPs.

Chapter 5 compares the bio-economies of South Africa and Taiwan. Taiwan is illustrative of the case of relying solely on foreign technologies, through various science and research collaborations with foreign entities to develop its local bioeconomy industries. South–South partnerships and cooperation with technologically advanced partners is recommended for South Africa. The government should ensure effective coordination and mobilisation of research efforts via PPPs, and joint research and entrepreneurship programmes. It should also ensure that innovation is not hindered due to unnecessary regulations and policies.

Chapter 6 compares the biotechnological research and product-development standing of India and South Africa. It argues that research collaborations, agreements and partnerships contributed to India's success in developing a strong technological and science base in biotechnology. It is critical for South Africa to speed up its technological learning process by encouraging innovation and learning through international collaborations and research partnerships rather than relying solely on developing domestic resources and local collaborations.

Chapter 7 compares stem-cell development research in South Africa and India and in both countries, stem-cell research forms part of the national biotechnology strategy. Compared to India, South Africa has weak industry and university linkages, and its stringent regulatory laws have stifled product innovation. As discussed earlier, public investment into research is not sufficient for a successful innovation system. Rather, certain cultural and institutional barriers need to be overcome in order to translate basic research into commercial success.

REFERENCES

Adelowo, CM, Ilori, MO, Siyanbola, WO & Oluwale, BA. 2015. 'Technological learning mechanisms in Nigeria's technology incubation centre'. *African Journal of Economic and Management Studies*, 6(1), pp 72–89.

Appelbaum, RP. 2016. *CNS Synthesis Report on IRG 2: Globalization and Nanotechnology: The Role of State Policy and International Collaboration.* CNS-UCSB, Santa Barbara, CA, July 2016.

Aydogan-Duda, N. ed. 2012. *Making It to the Forefront: Nanotechnology: A Developing Country Perspective* (Vol. 14). New York: Springer Science & Business Media.

Bell, M & Pavitt, K. 1995. 'The development of technological capabilities'. *Trade, Technology and International Competitiveness*, 22(4831), pp 69–101.

Beumer, K & Bhattacharya, S. 2013. 'Emerging technologies in India: Developments, debates and silences about nanotechnology'. *Science and Public Policy*, 40(5), pp 628–643.

Christensen, C. 1997. *The Innovator's Dilemma: When New Technologies Cause Great Firms to Fail.* Boston, Massachusetts: Harvard Business Review Press.

Cohen, WM & Levinthal, DA. 1990. 'Absorptive capacity: A new perspective on learning and innovation'. *Administrative Science Quarterly*, 35(1), pp 128–152.

Cozzens, S. 2012. 'The distinctive dynamics of nanotechnology in developing nations'. In Aydogan-Duda, N, ed. *Making It to the Forefront*. New York: Springer New York, pp 125–138.

Cozzens, SE. 2011. 'Introduction to the special issue on "Distributional consequences of emerging technologies"'. *Technological Forecasting and Social Change*. Available at: www.dx.doi.org/10.1016/j.techfore.2010.09.009 (accessed in February 2016).

Fagerberg, J. 2010. 'Innovation: A guide to the literature'. In Fagerberg, J, Mowery, DC & Nelson, R, eds. *The Oxford Handbook of Innovation*. Oxford, UK: Oxford University Press.

Gaskell, G & Bauer, MW. eds. 2001. *Biotechnology 1996–2000: The Years of Controversy*. London: Science Museum.

Gastrow, M, Roberts, B, Reddy, V & Ismail, S. 2016. 'Public perceptions of biotechnology in South Africa'. Available at: www.pub.ac.za (accessed on 14 February 2016).

Goven, J & Pavone, V. 2015. 'The bio-economy as political project: A polanyian analysis'. *Science, Technology & Human Values*, 40(3), pp 302–337.

Invernizzi, N & Foladori, G. 2005. 'Nanotechnology and the developing world: Will nanotechnology overcome poverty or widen disparities'. *Nanotechnology, Law & Business*, 2, pp 294–303.

Jasanoff, S. 2015. 'Imagined and invented worlds'. In Jasanoff, S & Kim, S, eds. *Dreamscapes of Modernity: Sociotechnical Imaginaries and the Fabrication of Power*. Chicago: University of Chicago Press.

Jasanoff, S & Kim, SH. 2009. 'Containing the atom: Sociotechnical imaginaries and nuclear power in the United States and South Korea'. *Minerva*, 47(2), p 119.

Kraemer-Mbula, E & Wamae, W. eds. 2010. *Innovation and the Development Agenda*. Canada: OECD Publishing.

Lema, R, Johnson, B, Andersen, AD, Lundvall, BÅ & Chaudhary, A. 2014. *Low-Carbon Innovation and Development*. Aalborg: Aalborg University Press.

Freeman, C & Lundvall, BA. 1988. *Small Countries Facing the Technological Revolution*. London: Frances Pinter Publishers Ltd.

Lundvall, BÅ. ed. 1992. *National Innovation System: Towards a Theory of Innovation and Interactive Learning*. London: Pinter Publishers.

Motoyama, Y, Appelbaum, R & Parker, R. 2011. 'The National Nanotechnology Initiative: Federal support for science and technology, or hidden industrial policy?'. *Technology in Society*, 33(1), pp 109–118.

Marchant, GE & Wallach, W. 2015. 'Coordinating technology governance'. *Issues in Science and Technology*, 31(4), pp 43–50.

Mazzucato, M. 2011. 'The entrepreneurial state'. *Demos*, London. Available at: www.demos.co.uk/files/Entrepreneurial_State_-_web.pdf. (accessed in February 2016).

Mazzucato, M. 2013a. *The Entrepreneurial State: Debunking the Public vs. Private Myth in Risk and Innovation*. London: Anthem Press.

Mytelka, L. 2004. 'Catching up in new wave technologies'. *Oxford Development Studies*, 32(3), pp 389–405.

OECD, 1997. National Innovation Systems. Available at: www.oecd.org/science/inno/2101733.pdf (accessed on 5 February 2016).

Page, J. 2005. 'Strategies for pro-poor growth: Pro-poor, pro-growth or both?' Paper presented at African Development and Poverty Reduction: The Macro-Micro Linkage, the DPRU/TIPS Forum, hosted in association with Cornell University, Cape Town, 13–15 October 2004.

Tegart, G. 2006. 'Critical issues in the commercialization of nanotechnologies'. *Innovation*, 8(4–5), pp 338–347.

Sandler, R. 2009. 'Nanotechnology: The social and ethical issues'. Available at: www.pewtrusts.org/~/media/legacy/uploadedfiles/phg/content_level_pages/reports/nanofinalpdf.pdf (accessed on 12 February 2016).

Van Horn, C & Fichtner, A. 2008. 'The workforce needs of companies engaged in nanotechnology research in Arizona'. The Center for Nanotechnology in Society, Arizona State University. Available at: www.cns.asu.edu/cnslibrary/year (accessed on 12 February 2016).

Teece, DJ. 1996. 'Firm organization, industrial structure, and technological innovation'. *Journal of Economic Behavior & Organization*, 31(2), pp 193–224.

Williams, R & Edge, D. 1996. 'The social shaping of technology'. *Research Policy*, 25(6), pp 865–899.

ONE

The advancement of nanotechnology

A sustainable development or an untenable vision?

ZAMANZIMA MAZIBUKO

THE EVOLUTION OF NANOTECHNOLOGY

NANOTECHNOLOGY ALLOWS for the manipulation of matter at the level of individual atoms and molecules which results in the construction of materials, devices and systems with new characteristics and applications made possible by their small structure (Staggers et al, 2008). The term nanotechnology is derived from the prefix 'nano' in nanometer (nm), which is one billionth of a metre and is approximately the width of 10 hydrogen atoms. A nanoparticle is defined as a collection of approximately $10–10^5$ atoms attached to each other with a radius between 1 and 100 nm (Bhushan, 2007). The main expressed motive for developing nanotechnology is to accelerate progress towards better healthcare, increase productivity and sustainable development.

Nanotechnology controls structures to atomic precision where every atom is positioned accordingly for the most favourable function of the nano-structure. Such technology has enormous potential: it could result in the use of fewer resources and energy and a reduction in the waste produced from manufacturing, as well as the development of new methods to convert energy and filter water, for example.

The concept of employing atomic precision in technology was first mentioned in 1959 at the California Institute of Technology by physicist and Nobel Laureate Professor Richard Feynman in his visionary talk, 'There's plenty of room at the bottom' (Feynman, 1960). In this talk Feynman disclosed his high-tech vision of miniaturisation of extreme proportions. He proposed that if atoms could be rearranged in any way we wished, while still being consistent with the laws of physics and chemistry, many devices could function at extremely efficient levels. He gave an example of computers and how they filled up an entire room because of their large size and yet were not able to store enough information to, for instance, select a method of analysis which would be better than the one it was programmed to use. He also pointed out that there was no machine that could recognise a face that it had been shown previously, if the picture was not exactly the same as before. To make a device that had these features, among others, would require a large amount of material, an excessive amount of heat and power, and the machine would still not perform at a desirable speed. Feynman anticipated that faster computers would need to be made extraordinarily smaller. He believed was this was highly possible as there was no law in physics that disputed this and that there was 'plenty of room to make them smaller' (Feynman, 1960). Although miniaturisation is a concept that had been perceived as being around for a very long time, the meticulous manipulation of the atomic structure that Feynman described was considered innovative.

The term 'nanotechnology' was, however, first proposed by Tokyo Science University Professor Norio Taniguchi in 1974. Taniguchi described nanotechnology as the 'processes of separation, consolidation, and deformation of materials by one atom or one molecule' (Taniguchi, 1974).

Twenty-seven years after the talk by Feynman, a book by K Eric Drexler titled *Engines of Creation* (Drexler, 1986) used the term nanotechnology to describe Feynman's vision and made it a more attractive and popular concept. Drexler predicted the endless possibilities that could potentially flow from the use of nanotechnology; the manufacturing of assembly machines smaller than living cells and the production of materials that are more durable yet are on a miniature scale. Drexler envisaged nanotechnology to be able to improve spacecrafts, repair living cells, heal diseases and allow humans to have stronger and faster bodies. He anticipated that nanotechnology would facilitate the conception of molecular machines that were so small and therefore so efficient that they could not only create materials capable of transforming our physical environment, but also advance the activities in that environment. These molecular machines or 'nanorobots' would be programmed to accomplish atomic precision, placing each atom into a specific arrangement, as per Feynman's vision. Drexler later referred to the vision that included these molecular machines as molecular manufacturing which he stated was 'a process of construction based on atom-by-atom control of product structures which may use assemblers (or more specialised mechanisms) to guide a sequence of chemical reactions' (Drexler, 2003).

Drexler saw nanotechnology as the solution to most, if not all, of the problems faced by the human race and predicted it would create a whole new world. Drexler's forecasts about the future role of nanotechnology laid the basis for the direction research inevitably took, essentially, technological determinism. Nanotechnology's foundation is pinned on the future more than it is on the present. It pushes the limits of human agency, but without societal buy-in and absorption into society advances cannot easily be made. The view that Drexler and Feynman held of nanotechnology has been classified as molecular nanotechnology (MNT), which differentiates it from nanotechnology that does not consider atomic precision. The promise that MNT has presented has attracted some scepticism and criticism. The science community distinguishes Feynman's vision from Drexler's, claiming that Drexler's idea is farfetched and is not technologically feasible.

The spokesperson for the US National Nanotechnology Initiative, Professor Richard Smalley, has been one of the biggest sceptics of the Feynman vision of nanotechnology. He was cited as dismissing Feynman's vision, stating that it is doubtful that MNT would be possible without 'magic fingers' to assemble devices by placing atoms in specific positions with precision (Smalley, 2001). Drexler's vision of nanotechnology was a long-term concept which would only produce tangible results decades after research commenced. Conversely, the focal point of the development of nanotechnology by the science community has deviated slightly to more attainable and instant targets; targets that do not include the creation of molecular machines in order to produce nanodevices. This deviation has resulted in the inclusion of some technologies under the umbrella definition of nanotechnology which would have been excluded under Drexler's conception. He has, however, claimed that molecular machines are an extension of Feynman's vision and not a deviation as reported (Drexler 2003).

The appeal of nanotechnology, especially to society, was mostly what Drexler envisioned and reported on. However, scientists and technologists think it is an impossible target especially with the time frame set by funders. Scientists and technologists currently use the term nanotechnology to refer to an applied science in which a material of nanoscale size exhibits characteristics that are different from the bulk material. A material that is reduced to a size below 100 nm shows distinct changes in properties (Lane & Kalil, 2005). These characteristics could be a difference in tolerance to temperature and pressure, conductivity, strength, elasticity and reactivity. The change in how the material reacts when it is at a much smaller scale allows nanotechnology to produce faster, cheaper, lighter, safer, cleaner and more defined solutions.

Scale-based definitions of nanotechnology also incorporate existing techniques and processes, but within a much smaller range. Descriptions of this nature apply to several companies that manufacture products using reactions at a small scale, such as catalyst production. These companies define their production as nanotechnology, thus making nanotechnology a much more expansive technology than initially perceived. Definitions of nanotechnology, some determined by government organisations, have evidently expanded over the years. Essentially, nanotechnology was initially designated solely as technology. However, at present commercial aspects are included in the definition of nanotechnology, and society has had an impact on the trajectory of the new technology.

ANCIENT NANOTECHNOLOGY: HOW FAR BACK DOES NANOTECHNOLOGY ACTUALLY GO?

Nanotechnology is not an entirely new notion. Chemistry, for example, is concerned with the organising of atoms into larger molecules, which can be combined to form polymers. Biology demonstrates atomic precision in the key molecules crucial to life and this can be seen in a cell which is made up of an accurate arrangement of atoms that form protein molecules that fold their long chains into specific molecules that have particular functions. These molecules are able to form large complexes of molecules that combine to form subcellular components in accordance with the information encoded in their precise sequences.

Aside from natural nanoscale reactions, chemical synthesis of nanoparticles has existed for many years. Some of the earliest evidence of the use of nanotechnology dates back more than 4000 years ago in Africa where ancient Egyptians initiated lead-based chemistry for cosmetic applications such as hair dyes (Walter et al, 2006) and black eye makeup (which was used to treat or prevent eye illnesses) (Tapsoba et al, 2010; Loyson, 2011). For hair dyes, a blend of lead oxide (PbO) and slaked lime (lime mixed with water to produce calcium hydroxide ($Ca(OH)_2$)) (Dei & Salvadori, 2006), with a small amount of water to form a paste, was applied on the hair. The deposition of galena/lead sulphide (PbS) crystals throughout the chemical reaction results in the blackening of the hair. The sulphur contained in the reaction originates from the amino acids of hair keratins and the lead is present in the paste that precipitates on the hair shafts. Lead-based chemistry has been shown to result in the formation of galena nanocrystals which are approximately 5 nm in size and have similar physical features to PbS quantum dots synthesised by modern techniques (Walter et al, 2006).

Moreover, clay minerals, which make up elements of soil particles smaller than 2 μm, were used as natural nanomaterials with many applications, including bleaching wool and clothes and, further, to remove oil from clothes (Rytwo, 2008). This specific application of clay is reported to date from 5000 BC in Cyprus (Rytwo, 2008). In China, between the 6th and 7th century AD, kaolin (a distinct, fine plastic clay) was used as a raw material to manufacture porcelain with a diameter of less than 0.4 mm (Yanyi, 1987; Rytwo, 2008). Clay was also used for medicinal purposes across the globe by people near the Dead Sea (Essenians), people in Africa, South America, Australia and elsewhere, who supplemented their diets with clay and also used it to assist in healing. In addition, clay was used worldwide in cosmetic applications (Rytwo, 2008).

In another example, the Maya (indigenous people of Mesoamerica) produced Maya blue, which is the bright turquoise colour seen in Mayan artefacts (Encyclopædia Britannica, 2016) first produced approximately in the 8th century AD (Chiari et al, 2008). The Mayans were able to create a crucial technique that bound indigo dye to a clay mineral substrate. Maya blue is able to resist potent nitric acid, alkali and organic solvents and still maintain its colour as well as withstand years of exposure to humid conditions (Chiari et al, 2008). Modern technology has been able to illuminate how the Mayans were able to create this technology. The presence of palygorskite (clay mineral) in Maya blue was observed using X-ray powder diffraction (XRPD), followed by the use of infrared spectroscopy to detect indigo (found in the plant *Indigofera suffruticosa*), thus confirming that Maya blue is a complex of the two elements. It was established that the mixture of indigo and palygorskite had to be heated to 100°C to produce Maya blue (Chiari et al, 2008). An analysis which drew on transmission electron microscopy (TEM) proposed that nanoscale iron (Fe), titanium (Ti) and manganese (Mn) impurities found in Maya blue samples may affect its appearance (Chiari et al, 2008).

Furthermore, gold and silver nanoparticles have been found on colourful Roman glass cups manufactured in the 4th century AD (Freestone et al, 2007). The Lycurgus cup, an elaborate and sophisticated Roman vessel, is green in reflected light but turns red in transmitted light. The presence of minute amounts of gold and silver were observed using microanalysis. Similar to Maya blue, the mere presence of certain elements is not enough to produce these unique characteristics in the colour. TEM technology revealed the presence of miniature metal particles (50–100 nm), while analysis by XRPD confirmed the particles to be a nanoscaled silver-gold complex, with a ratio of silver to gold of about 7:3 and an additional 10% copper (Freestone et al, 2007).

There are several other examples of metal nanoparticles being used to create unique colour features in ancient objects. Churches built in Rome between the 4th and 20th century AD are adorned with mosaic glass tiles, some of which are opaque-pinkish in colour for faces, hands and feet. The presence of colloidal gold nanoparticles between 10 and 35 ppm (parts per million) gives these tiles their colour (Veritã & Santopadre, 2010). However, between the 16th and 13th century BC, Egyptian glassmakers were already using nanoparticles to give glass high technical and aesthetic qualities. Egyptian glassmakers opacified opaque white, blue and turquoise glasses by using calcium antimonite crystals diffused in a vitreous mixture (Lahlil et al, 2010). This was done by formulating calcium antimonite opacifiers before incorporating these into a glass. In a study of Egyptian opaque glasses, TEM technology has revealed that these opacifiers were nanocrystals (Lahlil et al, 2010).

Despite the fact that nanotechnology has evidently existed for a very long time, it was only in the late 20th century that the hype around it was ignited, leading to a surge of research and intensified investigation as the field was labelled an emerging technology. One of the most apparent reasons for this extreme delay has been the lack of experimental equipment and techniques to conduct research at the nanoscale. The scanning probe microscopy, for example, was only established in 1981, with the invention of the scanning tunnelling microscope (The Center for Probing the Nanoscale).

Some of the numerous discoveries and inventions that have enabled the development of nanotechnology over the years are listed in chronological order below:

- *1981:* The *scanning tunnelling microscope* was invented by Gerd Binnig and Heinrich Rohrer, allowing scientists to observe singular atoms for the first time (Binnig & Rohrer, 1987).
- *1981:* A Ekimov and A Onushchenko uncovered *nanocrystalline, semiconducting quantum dots in a glass matrix* followed by the investigation of their electronic and optical properties (Ekimov & Onushchenko, 1981).

- *1985:* The *buckyball, or Buckminsterfullerene (C60)*, a small spherical carbon molecule, was discovered by Rice University researchers Harold Kroto, Sean O'Brien, Robert Curl and Richard Smalley (Bhatnagar & Goel, 2014; Chemical Heritage Foundation, 2015).
- *1985:* Louis Brus discovered *colloidal semiconductor nanocrystals (quantum dots)* (Bhatnagar & Goel, 2014).
- *1986:* The *atomic force microscope* was invented by Gerd Binnig, Calvin Quate and Christoph Gerber. It is able to view and manipulate matter at a nanoscale, including measurement of a range of forces fundamental to nanomaterials (Dartmouth Undergraduate Journal of Science, 2009).
- *1991:* The discovery of the *carbon nanotube,* which displays unique characteristics with regards to its strength, electrical and thermal conductivity, was credited to Sumio Iijima (Iijima, 2002).
- *1992:* CT Kresge and colleagues discovered the *nanosized catalytic materials MCM-41* and *MCM-48.* These materials are currently employed in refining crude oil as well as for water treatment and drug delivery (Krege et al, 1992).
- *1993:* A technique for *controlled formulation of nanocrystals* was invented by Moungi Bawendi which led to high-efficiency conversion of light to electricity at atomic scale using semiconducting materials among other processes (Nanotechnology Timeline).
- *1999:* The *dip-pen nanolithography (DPN)* was invented (Piner et al, 1999).
- *2003:* Naomi Halas, Jennifer West, Rebekah Drezek and Renata Pasqualin created *gold nanoshells*, which can absorb near-infrared light, and are important in the non-invasive detection, diagnosis and treatment of breast cancer (Gupta, 2014).
- *2007:* Angela Belcher and colleagues used a harmless virus to build a *lithium-ion battery* by means of an economical and environmentally friendly method (Gupta, 2014).
- *2009–2010:* Nadrian Seeman and colleagues developed a number of *DNA-like robotic nanoscale assembly devices* (Gu et al, 2010).
- *2013:* Stanford researchers created the first carbon nanotube computer (Gupta, 2014).

There are currently several nanotechnology commodities available on the market. Sunscreens have nanoparticles to make them transparent; nanomembranes have been employed for water purification and manufacturers use chemical processing by catalysis. Computer memory is built using nanotubes to store information. Faster computer processors, cosmetics, antibiotic bandages and stain-resistant clothing all have nanotechnology incorporated, resulting in vast improvements in their efficiency, cost, appearance and weight. Other nanotechnologies are still being investigated; some are concepts that still require years to be developed.

In 2014, global nanotechnology funding from both the public and private sector was estimated by an emerging technologies consulting firm, Lux Research, to be approximately $18.5 billion in 2012 (Lux Research, 2014). The introduction into the market of products incorporating nanotechnology has seen these global investments begin to yield highly anticipated revenue. There has been an exponential growth in global sales from 2004 to 2015 according to Lux Research. This exponential growth is in all probability because nanotechnology can be applied to products in any sector. It must be noted, however, that nanomaterials contribute to less than 0.5% of the global sales, whereas nano-enabled products yield the most revenue followed by nano-intermediates. Nano-intermediates are intermediate products that have incorporated nano-sized materials. These products, for example, fabrics and memory chips, are usually not yet targeted at the final user.

In 2014, Lux Research estimated that by 2018 products incorporating nanotechnology will yield $4.4 trillion in revenues. The claim by scientists as well as business is that with the increased integration of nanotechnology with biotechnology, cognitive sciences and information technology, more radical inventions will surface that will revolutionise medicine, agriculture, the textile industry, energy technologies and national security (Sargent, 2014).

It is evident that government support for the development of nanotechnology plays a key role in ensuring the viability and success of this emerging technology. In the next section, the United States of America, as one of the leading countries in nanotechnology development, is examined and the US government's contribution towards the growth of nanotechnology in the country is explored. This is then compared with South Africa's nanotechnology efforts and the extent of government support in a country that still needs to catch up to developed countries.

US GOVERNMENT SUPPORT FOR NANOTECHNOLOGY

The USA has been the largest contributor to the advancement of nanotechnology, through government involvement via the introduction of various programmes. The Interagency Working Group on Nanotechnology (IWGN) was established under the National Science and Technology Council to review nanoscience and nanotechnology and predict potential developments (Nanotechnology Timeline). The IWGN's findings, which were presented in a report titled 'Nanotechnology research directions: Vision for the next decade' (1999), led to the creation of the US National Nanotechnology Initiative (NNI) (Roco et al, 1999). This was established in 2000, by President Bill Clinton. The NNI is a US research and development initiative that combines the varied expertise needed to accelerate the development of nanotechnology. It is made up of 20 government departments and autonomous agencies that ensure that nanotechnology research and development occurs in tertiary institutions, government and industry laboratories across the United States. The NNI has since its inception conducted various investigations, public workshops and dialogues to produce strategic plans and progress reviews.

The 21st Century Nanotechnology Research and Development Act (2003) was passed in the USA to facilitate research programmes for nanoscience, nano-engineering and nanotechnology. The Act provided a statutory platform for a National Nanotechnology Research Programme with a National Nanotechnology Coordination Office. The Act also allowed for a research programme that would address the ethical, legal and environmental issues associated with nanotechnology. It required that the funding and establishment of research centres be done on a competitive basis. Furthermore, the Act extended to research in biotechnology and applications in various fields.

It is evident that the US government is extensively involved in the development of nanotechnology in the country. Government funding allows for the exponential development of new technologies and, therefore, patents and published papers follow. According to a survey taken in 2008, there were approximately 400 000 global workers in the nanotechnology sector, with an estimated 150 000 of those in the United States (Roco et al, 2011).

NANOTECHNOLOGY IN SOUTH AFRICA

In South Africa, nanotechnology has been included in the national strategy since 1996 when the White Paper on Science and Technology was published. Ten years later, the National Nanotechnology Strategy (NNS) was crafted. The South African government regards science and technology as a way to accelerate development by moving the country towards a knowledge economy, with improved employment rates and reduced poverty. The Department of Science and Technology has been on a mission to equip the country with the human capital and capacity for the innovation required to compete globally. Nanotechnology is viewed as a tool that will achieve sustainable economic growth in an environmentally sustainable manner. The promise of nanotechnology, namely to provide smaller, cheaper products which require fewer raw materials and energy, seems particularly ideal for a developing country.

Various organisations in South Africa, ranging from businesses to science councils and institutions of higher education have been actively involved in contributing to the development of nanotechnology in the country. The NNS focuses on six areas, namely, water, energy, health, chemical and bio-processing, mining and minerals, and advanced materials and manufacturing. Researchers in the country have successfully used nanotechnology to improve existing technologies and continue to investigate ways of doing so. The Council for Scientific and Industrial Research (CSIR), for example, has been researching using nanotechnology in the treatment of TB. A number of universities have partnered with science councils and industrial companies to research and produce innovative nanotechnology devices. The Department of Science and Technology established the national facility Mintek Nanotechnology Innovation Centre (NIC) in 2007. The NIC is constituted of three science councils namely Mintek, Medical Research Council (MRC) and Water Research Commission (WRC) and three universities: University of Johannesburg (UJ), Rhodes University (RU) and the University of the Western Cape (UWC). The Department of Applied Chemistry at UJ, for instance, is in collaboration with the NIC to develop water treatment solutions using nanomaterials. The National Centre for Nano-Structured Materials (NCNSM), also a national facility, was jointly initiated by the DST and the CSIR in 2007 and collaborates with the University of Cape Town, University of the Free State, University of the Witwatersrand, University of Western Cape, University of Pretoria, University of Zululand, Nelson Mandela University, North-West University and iThemba Laboratories (a national research facility managed by the National Research Foundation (NRF). Health, water and energy are considered the areas that need the most attention as they will have the greatest impact on society. Nanotechnology initiatives are, therefore, focused on these areas.

This widening use of nanotechnologies has raised questions about the risks associated with the technology. This is a concern that scientists rarely engage with while formulating nanoproducts in the laboratory. This aspect of nanotechnology is briefly discussed in the next section.

THE RISKS AND UNCERTAINTIES OF NANOTECHNOLOGY

Currently, as with most emerging technologies, the science and progress of nanotechnology is far ahead of social understanding, ethics and policies. Concerns have been raised about insufficient knowledge of the undesirable properties of this emerging technology, including health and environmental risks as well as a displaced workforce.

Much of the literature about nanotechnology notes that it is vital for the wellbeing of society, as long as it is pursued in a socially responsible manner. Nanoparticles in drug delivery systems, for example, are able to evade natural defences such as the blood-brain barrier, penetrate cells in organisms, and travel to targeted organs and tissues to deliver drugs. This is highly beneficial. But, in contrast, the very same impressive features of nanoparticles could prove to be detrimental to the human biological system by allowing nanoparticles to remain in the body indefinitely, for instance.

The regulation of nanotechnology is undoubtedly an urgent matter. However, the difficulty with defining the risks of nanotechnology, and therefore of establishing guidelines, has been the fact that the products and the manufacturing thereof are merged with various other technologies. This combination of technologies makes it challenging to classify products and their proposed applications.

The food industry provides an example of the difficulty in providing concise and rational regulations. There is potential to use nanotechnology to improve the packaging of food and to enhance its nutritional properties without changing the taste or quality of the food. As with drug delivery, concerns about using nanotechnology in food relates to their possible harmful effects in the human body. Recently, products sold in the European Union (EU) have been required to be labelled with any nanoproducts used in the food. Food manufacturers are concerned that the size range prescribed for nanotechnology by the EU is problematic as it applies to 50% of materials in an ingredient. Labelling a product would thus result in an overload of information on the packaging, which would alarm consumers. Additionally, Dr Gareth Cave of Nottingham Trent University has stated that nanomaterials are already present in food, like flour for example, at the nanoscale and would thus, according to the EU, be 100% nano (Food Processing Technology, 2013).

Such confusion and discrepancies will continue making it difficult for legislators to put in place defined regulations for nanotechnology. South Africa has established an ethics committee to ensure ethical values are applied in the development of nanotechnology. The National Research Plan (NRP) has been established to assess the risk of nanomaterials and supplies information obtained from scientific data. This data includes identification of hazardous material, exposure assessment and risk characterisation. There is also the Nanotechnology Public Engagement Programme (NPEP) which encourages trustworthy and factual provision of information about nanotechnology through communication, awareness and educational programmes.The questions around nanotechnology extend to whether it is consistent with sustainable development, as well as regulatory concerns.

IS NANOTECHNOLOGY COMPATIBLE WITH SUSTAINABLE DEVELOPMENT?

Sustainable development has not always had a clear definition and has taken different forms over the years (Lélé, 1991; Kates et al, 2005). The concept was developed during a time when an increasing awareness of the damage to the environment and lack of concern for social issues was persistent (Dempsey et al, 2009). Currently, the commonly accepted or quoted definition for sustainable development, cited below, is derived from a document titled *Our Common Future*, also known as the Brundtland Report (United Nations, 1987: 204):

> *Sustainable development is development that meets the needs of the present without compromising the ability of future generations to meet their own needs. It contains within it two key concepts:*
> - *the concept of needs, in particular the essential needs of the world's poor, to which overriding priority should be given; and*
> - *the idea of limitations imposed by the state of technology and social organization on the environment's ability to meet present and future needs.*

The environmental and social sustainability of an emerging technology is always closely examined as there is, understandably, inadequate knowledge about it. Sustainability is an important aspect to consider when implementing a fairly new technology, ensuring that the requirements of the present as well as future generations are met.

Nanotechnology is a highly hyped-up term that instantly draws the attention of investors, whether private companies or government, and high-tech enthusiasts alike. It is, therefore, not surprising that there has been some apprehension around whether nanotechnology is consistent with sustainable development.

There are concerns about whether nanotechnology is following the same route taken by the technology used to create genetically modified organisms (GMO), with possibly the same vast public outcries about safety and essentiality (ETC Group, 2003). Furthermore, there were several promises that GMO would help in the advancement of developing countries. However, weaknesses in the regulatory environment are reflected in the fact that there are reports of contaminated corn in Oaxaca, Mexico (SciDev.Net, 2002a). In response to the hype around these technologies, Invernizzi and Foladori (2005) go to the other extreme, arguing that social mobilisation and traditional medicine are more effective than nanotechnology in meeting the needs of developing countries. They point out how China and Vietnam were able to considerably reduce malaria-related deaths without the use of nanotechnology.

Nanotechnology is being proclaimed as the answer to minimising our ecological footprint while advancing development. There are also claims that it facilitates economic growth through the development of fresh and innovative products and the consequent emergence of new markets. There is, however, insufficient evidence to fully support these statements and most of the proclamations are still quite theoretical. There are concerns that the claims made for nanotechnology do not match reality, or that they are highly exaggerated. It seems that in nanotechnology development, especially the initial stages, the environmental risks were underestimated or simply overlooked.

When assessing the sustainability of a nanoproduct, it is essential to understand how ecosystems and society will be affected by the entire life span of the product, i.e. its life cycle (Figure 1.1). Researchers started implementing all-inclusive assessment tools like Life Cycle Assessment (LCA) to assess the sustainability of products (Meyer et al, 2009). The important life cycle stages in a product's existence are:

- **Raw materials extraction**: Activities related to the acquisition of natural resources, including mining non-renewable material, harvesting biomass and transporting raw materials to processing facilities.

- **Materials processing**: Processing of natural resources by reaction, separation, purification and alteration steps in preparation for the manufacturing stage; and transporting processed materials to product manufacturing facilities.
- **Product manufacture:** Manufacturing of product and transporting to consumers.
- **Product use:** Use and maintenance activities associated with the product by the consumer.
- **End-of-life disposition:** Disposal of the product after its life span, which may include transportation, recycling, disposal, or incineration (Eason et al, 2011: 9).

Figure 1.1: The life cycle of a product

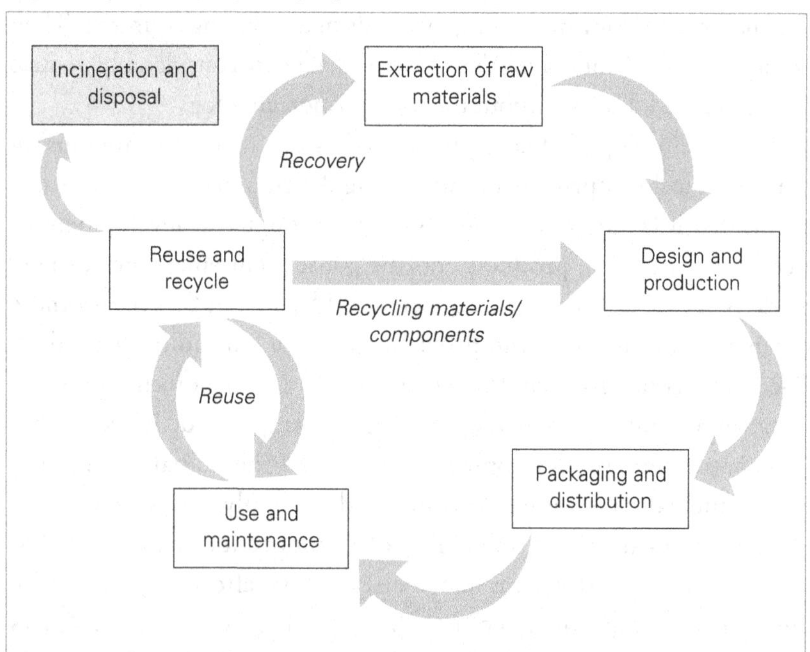

All stages of the life cycle consume resources and result in a waste product. Yet, often when examples are cited of how nanotechnology can benefit society, it is not the entire life cycle of the nanoproduct that is considered, but only the product use stage. This means that the greenhouse gas emissions and the waste generated at other stages of the product's life cycle are disregarded. It is highly possible that when production and disposal stages are taken into account, the net impact on the environment is negative. For example, carbon nanotubes, which are a durable and firm material, are estimated to require a total of 0.1-1 TJ/kg (terajoules per kilogram) of energy, when accounting for the entire manufacturing process (Gutowski et al, 2010).

The excessive amount of energy used when manufacturing nanoproducts using carbon is also seen in the manufacturing of carbon nanofibers (CNF). Khanna and his colleagues (Khanna et al, 2008) implemented a life cycle energy analysis of CNF synthesis compared with traditional materials on an equal mass basis. Their preliminary results showed that life cycle energy requirements for CNF were 6–60 times higher when compared with aluminium, steel and polypropylene. Further assessments proposed that CNFs may have a more adverse environmental impact when compared to traditional materials per kg of product. In another example, Meyer and his team (Meyer et al, 2011) used the LCA to investigate the difference in environmental impact between a pair of socks (60.8 g) containing 10.8 mg of nanoscale silver and a typical pair. Their results indicated that there was an insignificant difference between the nano-enabled socks and the normal pair. This is ascribed to the small quantity of silver that was added to the pair of socks. However, even with that minute difference, gas phase processes for the production of nanoscale silver led to increased negative impact on the environment.

It has been reported, however, that the LCA method has not been fully developed for nanoproducts and that there are still some stages that are not considered during the assessments of various products (Hischier & Walser, 2012). According to Hischier and Walser (2012), more research and defining of the LCA is required in order to produce useful values for nanoproducts.

In this context, some researchers have been aiming to achieve 'green nanotechnology', which refers to alleviating risks in various aspects of nanotechnology development and use. This concept involves three main considerations:

a) improving the development of clean technologies that incorporate nanotechnology;
b) reducing possible environmental and human health risks related to the production and use of nanoproducts; and
c) promoting the replacement of existing products with new nanoproducts that are more environmentally friendly throughout their life cycles (Schmidt, 2007).

A significant number of researchers in developing countries, together with the United Nations, see the potential of nanotechnology in fields such as water purification, solar cell technology and healthcare (SciDev.Net, 2002b). They consider nanotechnology as being compatible with sustainable development. Thus far, nano-enabled innovations have been able to improve technological efficiencies in various sectors. These include drug-delivery systems for the treatment or management of complex and defiant diseases such as tuberculosis (Choudhary & Devi, 2015), HIV/AIDS (Das Neves et al, 2010), neurodegenerative disorders (Modi et al, 2010) and various cancers (Aliosmanoglu & Basaran, 2012). The increase in the pharmaceutical efficacy of approved drugs using nanotechnology will certainly be highly beneficial for developing countries. For instance, lipid vesicles or liposomes, encapsulating doxorubicin (Doxil), were developed and subsequently approved for the treatment of AIDS-associated Kaposi's sarcoma (Farokhzad & Langer, 2009).

Furthermore, large strides have been made in advancing water treatment in developing countries employing nanotechnology. Additionally, it was estimated that nanotechnology would create a considerable number of jobs in the global market by 2020 (Roco et al, 2011).

It is difficult to assess whether nanotechnology is environmentally and socially sustainable as there are no definitive measures of this. The claims of nanotechnology's splendour have not been backed up with concrete evidence of an ecological footprint which would support sustainability. The advances made thus far in nanotechnology focus only on technological innovation and economic benefits, while ignoring the environmental and human health risks. The evidence from the early development of some nanotechnologies suggests manufacturing these products demands a greater amount of energy than the manufacture of similar conventional products. There has not been a widespread investigation of the sustainability of nanotechnologies and this is an area that needs to be addressed. Societal and governmental involvement would help to reduce the gap between the development of nanotechnologies and nanoethics. The lack of a suitable and meticulous risk and life cycle analysis of nanotechnologies to align the commercialisation of these products with societal and environmental benefits could eventually have an overall detrimental effect on the health of citizens and the environment. A more comprehenisive treatment of these issues is undertaken in the next chapter of this book.

CONCLUSION

Nanotechnology, broadly defined, has a longer history than is commonly perceived, with its application traced back to ancient Egypt. Even with this enduring history, there is still a question about whether or not nanotechnology is sustainable. What is evident, however, is that government support, including robust policies and substantial investments yields growth in the development and utility of nanotechnology. Society's involvement in the development of emerging technologies is also crucial for these technologies to advance. A multidisciplinary approach to comprehending nanotechnology and drawing up holistic policies is, therefore, key to ensuring nanotechnology's holistic development for the benefit of humanity.

REFERENCES

Aliosmanoglu, A & Basaran, I. 2012. 'Nanotechnology in cancer treatment'. *Journal of Nanomedicine and Biotherapeutic Discovery*, 2(4), pp 1–3.

Bhatnagar, S & Goel, A. 2014. 'Evolution of nanotechnology'. *BioEvolution*, 1(3), pp 76–79.

Bhushan, B. 2007. 'Introduction to nanotechnology'. In Bhushan, B, ed.

Springer Handbook of Nanotechnology. 2nd ed. Berlin/New York: Springer, pp 1–12.

Binnig, G & Rohrer, H. 1987. 'Scanning tunnelling microscopy: From birth to adolescence'. *Reviews of Modern Physics*, 59(3), pp 615–625.

Center for Probing the Nanoscale. 2009. 'Scanning Probe Microscopy'. Available at: www.teachers.stanford.edu/activities/SPMReference/SPM Reference.pdf (accessed on 29 June 2015).

Chemical Heritage Foundation. 2015. 'Richard E. Smalley, Robert F. Curl, Jr., and Harold W. Kroto'. Available at: www.chemheritage.org/historical-profile/richard-smalley-robert-curl-harold-kroto (accessed on 5 July 2015).

Chiari, G, Giustetto, R, Druzik, J, Doehne, E & Ricchiardi, G. 2008. 'Pre-Columbian nanotechnology: Reconciling the mysteries of the Maya blue pigment', *Applied Physics A*, 90(1), pp 3–7.

Choudhary, S & Devi, VK. 2015. 'Potential of nanotechnology as a delivery platform against tuberculosis: Current research review'. *Journal of Controlled Release*, 202, pp 65–75.

CSIR website: www.csir.co.za (accessed on 29 June 2015).

Dartmouth Undergraduate Journal of Science. 2009. 'Atomic force microscopy: Revolutionizing the future of nano-scale imaging'. Available at: www.dujs.dartmouth.edu/2009/11/atomic-force-microscopy-revolutionizing-the-future-of-nano-scale-imaging/#.Wi9obVWWaM8 (accessed on 5 July 2015).

Das Neves, J, Amiji, MM, Bahia, MF & Sarmento, B. 2010. 'Nanotechnology-based systems for the treatment and prevention of HIV/AIDS'. *Advanced Drug Delivery Reviews*, 62(4–5), pp 458–477.

Dei, L & Salvadori, B. 2006, 'Nanotechnology in cultural heritage conservation: Nanometric slaked lime saves architectonic and artistic surfaces from decay'. *Journal of Cultural Heritage*, 7(2), pp 110–115.

Dempsey, N, Bramley, G, Power, S & Brown, C. 2009, 'The social dimension of sustainable development: Defining urban social sustainability'. *Sustainable Development.* Available at: www.doi:10.1002/sd.417 (accessed in December 2016).

Drexler KE. 2003. 'Nanotechnology: From Feynman to funding'. *Bulletin of Science, Technology & Society*, 24(1), pp 21–27.

Drexler KE. 1986. *Engines of Creation: The Coming Era of Nanotechnology*. New York: Doubleday.

Eason, T, Meyer, DE, Curran, M & Upadhyayula, VKK. 2011. 'Guidance to facilitate decisions for sustainable nanotechnology', U.S. Environmental Protection Agency, Washington, D.C., USA.

Ekimov, AI & Onushchenko, AA. 1981. 'Quantum size effect in three-dimensional microscopic semiconductor crystals'. *Journal of Experimental and Theoretical Physics Letters (JETP Letters)*, 34(6), pp 345–349.

Encyclopædia Britannica. 2016. 'Maya'. Encyclopædia Britannica, inc.

Available at: www.britannica.com/topic/Maya-people (accessed on 14 January 2016).

ETC Group. 2003. 'The big down: Atomtech – Technologies converging at the nanoscale'. Available at: www.etcgroup.org/content/big-down-0 (accessed in July 2015).

Farokhzad, OC & Langer, R. 2009. 'Impact of nanotechnology on drug delivery'. *American Chemical Society Journal*, 3(1), pp 16–20.

Feynman, R. 1960. 'There's plenty of room at the bottom'. *Caltech Engineering and Science,* 23(5), pp 22–36.

Food Processing Technology. 2013. 'Nanotechnology in food: Research, development and labelling'. Available at: www.foodprocessing-technology.com (accessed on 30 June 2015).

Freestone, I, Meeks, N, Sax, M & Higgitt, C. 2007. 'The Lycurgus cup – A Roman nanotechnology'. *Gold Bulletin*, 40(4), pp 270–277.

Gu, H, Chao, J, Xiao, S & Seeman, NC. 2010. 'A proximity-based programmable DNA nanoscale assembly line'. *Nature*, 465(7295), pp 202–206.

Gupta, SRN. 2014. 'Advances in molecular nanotechnology from premodern to modern era'. *International Journal of Materials Science and Engineering*, 2(2), pp 99–106.

Gutowski, TG, Liow, JYH & Sekulic, DP. 2010. 'Minimum exergy requirements for the manufacturing of carbon nanotubes', *Sustainable Systems and Technology*, IEEE International Symposium on Sustainable Systems and Technologies, Washington D.C., May 16–19, 2010.

Hischier, R & Walser, T. 2012, 'Life cycle assessment of engineered nanomaterials: State of the art and strategies to overcome existing gaps'. *Science of the Total Environment*, 425, pp 271–282.

Iijima, S. 2002. 'Carbon nanotubes: Past, present, and future'. *Physica B*, 323, pp 1–5.

International Risk Governance Council. 2006. White Paper on Nanotechnology Risk Governance. Available at: www.irgc.org/IMG/pdf/IRGC_white_paper_2_PDF_final_version-2.pdf (accessed on 30 May 2015).

Invernizzi, N & Foladori, G. 2005. 'Nanotechnology as a solution to the problems of developing countries?' Available at: www.archive.cspo.org/_old_ourlibrary/documents/NanoSolut.pdf (accessed on 30 May 2015).

Kates, RW, Parris, TM & Leiserowitz, AA. 2005. 'What is sustainable development?'. *Environment*, 47(3), pp 8–21.

Khanna, V, Bakshi, B & Lee, L. 2008. 'Carbon nanofiber production: Life cycle energy consumption and environmental impact', *Journal of Industrial Ecology*, 12(3), pp 394–410.

Kresge, CT, Leonowicz, ME, Roth, WJ, Vartuli, JC & Beck, JS. 1992. 'Ordered mesoporous molecular sieves synthesized by a liquid-crystal template mechanism'. *Nature*, 359(6397), pp 710–712.

Lahlil, S, Biron, I, Cotte, M, Susini, J & Menguy, N. 2010. 'Synthesis of calcium antimonate nano-crystals by the 18th dynasty Egyptian glassmakers', *Applied Physics A*, 98, pp 1–8.

Lane, N & Kalil, T. 2005. 'The National Nanotechnology Initiative: Present at the Creation'. *Issues in Science and Technology*, 21(4), pp 49–54.

Lélé, SM. 1991. 'Sustainable development: A critical review'. *World Development*, 19(6), pp 607–621.

Loyson, P. 2011. 'Chemistry in the time of the Pharaohs'. *Journal of Chemical Education*, 88(2), pp 146–150.

Lux Research. 2014. 'Nanotechnology update: Corporations up their spending as revenues for nano-enabled products increase'. Available at: www.portal.luxresearchinc.com/research/report_excerpt/16215 (accessed on 1 July 2015).

Meyer, D, Curran, M & Gonzalez, MA. 2009, 'An examination of existing data for the industrial manufacture and use of nanocomponents and their role in the life cycle impact of nanoproducts'. *Environmental Science & Technology*, 43(5), pp 1256–1263.

Meyer, D, Curran, M & Gonzalez, MA. 2011, 'An examination of silver nanoparticles in socks using screening-level life cycle assessment'. *Journal of Nanoparticle Research*, 13(1), pp 147–156.

Modi, G, Pillay, V & Choonara YE. 2010. 'Advances in the treatment of neurodegenerative disorders employing nanotechnology'. *Annals of the New York Academy of Sciences*, 1184(1), pp 154–172.

Nanotechnology Timeline. Available at: www.nano.gov/timeline (accessed on 8 July 2015).

National Nanotechnology Strategy. Available at: www.gov.za/sites/www.gov.za/files/DST_Nanotech_18012006_0.pdf (accessed on 26 May 2015).

Office of Legislative Policy and Analysis. Available at: www.olpa.od.nih.gov/legislation/108/publiclaws/nanotechnology.asp (accessed on 9 September 2015).

Piner, RD, Zhu, J, Xu, F, Hong, S & Mirkin, CA. 1999. '"Dip-Pen" Nanolithography'. *Science*, 283(5402), pp 661–663.

Roco, MC, Mirkin, CA & Hersam, MC. 2011. 'Nanotechnology research directions for societal needs in 2020: Summary of international study'. *Journal of Nanoparticle Research*, 13(3), pp 897–919

Roco, MC, Williams, S & Alivisatos, P. 1999. 'Nanotechnology research directions: Vision for nanotechnology in the next decade'. IWGN Workshop Report, U.S. National Science and Technology Council, Washington D.C.

Rytwo, G. 2008, 'Clay minerals as an ancient nanotechnology: Historical uses of clay organic interactions, and future possible perspectives'. *Macla*, 9, pp 15–17.

Sargent Jr, FJ. 2014. *Overview, Reauthorization, and Appropriations Issues*. The National Nanotechnology Initiative, Congressional Research Service, USA

Schmidt, KF. 2007, 'Green nanotechnology: It's easier than you think'. Project on emerging nanotechnologies, 4th symposium on nanotechnology and the environment, Washington D.C.

SciDev.Net. 2002a. 'Mexico confirms GM maize contamination'. Available at: www.scidev.net/global/gm/news/mexico-confirms-gm-maize-contami= nation.html (accessed on 30 May 2015).

SciDev.Net. 2002b. 'New technologies "central to sustainable development"'. Available at: www.scidev.net/global/capacity-building/news/new-technologies -central-to-sustainable-developme.html (accessed on 30 May 2015).

Senadheera, J. Bright Hub Engineering. 2009. *The Future of Nanotechnology*. Available at: www.brighthubengineering.com/manufacturing-technology /59283-the-future-of-nanotechnology/, (accessed on 3 June 2015).

Smalley, RE. 2001. 'Of chemistry, love, and nanobots'. *Scientific American*, 285 (September), pp 76–77.

Staggers, N, McCasky, T, Brazelton, N & Kennedy, R. 2008. 'Nanotechnology: The coming revolution and its implications for consumers, clinicians and informatics'. *Nursing Outlook*, 56(5), pp 268–274.

Taniguchi, N. 1974. 'On the basic concept of "nano-technology"'. Proceedings of the international conference on production engineering, Tokyo, Part II, Japan Society of Precision Engineering.

Tapsoba, I, Arbault, S, Walter, P & Amatore, C. 2010. 'Finding out Egyptian gods' secret using analytical chemistry: Biomedical properties of Egyptian black makeup revealed by amperometry at single cells'. *Analytical Chemistry*, 82, pp 457–460.

United Nations. 1987. *Our Common Future: Brundtland Report*. Oxford: Oxford University Press.

Veritä, M & Santopadre, P. 2010, 'Analysis of gold-colored ruby glass tesserae in Roman church mosaics of the fourth to 12th centuries'. *Journal of Glass Studies*, 52, pp 11–24.

Walter, P, Welcomme, E, Halle´got, P, Zaluzec, NJ, Deeb, C, Castaing, J, Veyssie`re, P, Bre´niaux, R, Le´ve^que, J & Tsoucaris, G. 2006. 'Early use of PbS nanotechnology for an ancient hair dyeing formula'. *Nano Letters*, 6(10), pp 2215–2219.

Yanyi, G. 1987. 'Raw materials for making porcelain and the characteristics of porcelain wares in north and south China in ancient times'. *Archaeometry*, 29(1), pp 3–19.

TWO

Nanoscience, nanotechnology, nanomaterials and nanotoxicology in South Africa

PULENG MATATIELE, NATASHA SANABRIA, MELISSA VETTEN & MARY GULUMIAN

INTRODUCTION

NANOSCIENCE REFERS TO THE STUDY of the phenomena and manipulation of materials at atomic, molecular and nanomolecular scales. Nanotechnology refers to the design, characterisation, production and application of structures, devices and systems by controlling shape and size at nanometre scale for practical application, in information technology, energy, environmental science, medicine, food safety and transportation (McNeil, 2005; Buzea et al, 2007; Shinde et al, 2012) and many other fields. According to the definition given in the technical report of the International Standards Organization (ISO), engineered nanomaterials (ENMs) encompass nano-sized objects with one or more external dimension in the nano-scale (ISO, 2010). ISO has also described a system by which wide range of nanomaterials can be classified, where ENMs may be distinguished by their shape as either nanoparticles (NPs) (all three dimensions in the nanoscale), nanofibres (two dimensions in the nanoscale, including nanowires, nanotubes and nanorods), or nanoplates (one dimension in the nanoscale). Nanomaterials may also be classified by similarities in chemical composition such as carbon-based, metals or metal oxides and organic nanomaterials.

The discipline of toxicology traditionally addresses adverse effects of chemicals to humans, animals and the environment, where the effects are correlated with the dose of the chemical. In contrast, the sub-discipline of particle toxicology addresses the adverse effects of particles and fibres (Elsaesser & Howard, 2012). Recently, nanotoxicology was proposed as a subdivision of particle toxicology and was further proposed to address the adverse health effects caused by ENMs (Donaldson et al, 2004). The toxicity of ENMs depends on various factors, including size, shape, aggregation, dissolution, chemical composition, crystallinity, surface charge, surface structure and the presence or absence of functional groups. Investigations into the potential toxicity of nanomaterials have, however, not kept pace with the rapidly growing nanotechnology, and our understanding of the adverse effects on human and environmental health lags behind development (Bernhardt et al, 2010). Given that the production volume of some ENMs is already exceeding thousands of tonnes, the potential health risks – whether actual or perceived – associated with the manufacture, distribution and use of ENMs must be balanced by the overall benefit that nanotechnology has to offer in various applications, including therapeutics and diagnostics.

The aim of this chapter is to review the available up-to-date information on the toxicity of ENMs and also summarise the current state of nanotechnology in South Africa. This review should foster a robust, public debate about nanotechnology and highlight the need for increased funding for research into the toxic effects of ENMs. It should serve to influence legislators to work to control and limit the widespread proliferation of consumer products containing ENM nanotechnologies, until a robust regulatory programme is in place.

Such a programme would aim to make nano-research activities comprehensible and be performed in a transparent manner, be accountable, be safe and sustainable, as well as non-threatening to human health and the environment.

TOXICOLOGICAL STUDIES ON POTENTIAL RISKS TO BIOLOGICAL SYSTEMS

The novel physicochemical properties of nanomaterials allow new applications in biomedical, optical and electronic fields. For example, their size-related properties and functions (with increased surface area to volume dimension ratio), make them fall below the critical wavelength of light, which results in them being transparent and suitable for packaging, cosmetics and coatings. In addition, the surfaces can be tailored as required to increase their dispersion in solvents. ENMs are also used in sporting goods, tyres, stain-resistant clothing, sunscreens, toothpaste, food additives, coatings to dissipate and minimise static electricity in fuel lines and hard disk handling trays. They are also found in electrostatically paintable car exterior components, flame-retardant fillers for plastics, as well as field emitter sources in flat panel displays. Other applications of ENMs include electronics, engineering and the fields of solar cells or catalysis, molecular encapsulation for energy and light harvesting photo activity, and rheology (Stewart & Fox, 1996; Hofkens et al, 2000). For example, the carbon-based ENMs, single-walled, double-walled or multi-walled carbon nanotubes (SWCNTs, DWCNTs and MWCNTs respectively) are used for electronics, optics and applications as additives to various structural materials including baseball bats, golf clubs, car parts or steel (Gullapalli & Wong, 2011).

Dendrimers are currently being investigated for biomedical applications (Abbasi et al, 2014) including medical imaging, tissue-targeted therapy, drug delivery and gene transfection (Hussain et al, 2004), as carriers for penicillin (Yang & Lopina, 2003), and in anticancer therapy (Quintana et al, 2002). Quantum dots, which are markers in biological cellular imaging in living cells and tissues, are used in exploratory medical diagnostics and therapeutics, as well as in self-assembly of nano-electronic structures. Metal oxides including titanium dioxide (TiO_2), zinc oxide (ZnO), and iron oxides are used as chemical polishing agents from semi-conductor wafers, as scratch resistant coatings for glass, or as cosmetics and sunscreens. The percentage distribution of ENMs for various applications recorded in 2007 included 53% for various chemicals, 34% for semiconductors, but only 7% for electronics, 3% for aerospace and defence, 2% for pharmaceuticals and healthcare, and finally 1% for the automotive industry (Vlachogianni et al, 2013). The TiO2 nanoparticle production was reported to be 50 400 tonnes in 2010 and has been predicted to increase to 201 500 tonnes by 2015 (Binh et al, 2014).

Hazard identification

The same desirable physicochemical characteristics of ENMs are reasons for concern, especially when the production levels of some types of ENMs ranges from well-established multi-tonne production per year of carbon black (for car tyres), to microgram quantities of fluorescent quantum dots (Barroso, 2011). ENM hazard identification and the subsequent risk assessment are of paramount importance with the understanding that ENMs are not just one single material. This makes it more challenging to assess their toxicity.

Size is understandably the first physicochemical characteristic that affects the toxicity of ENMs. A recent review has summarised the correlation between physicochemical properties and nanomaterial toxicity, where the critical size of ENMs were seen to be 30 nm and smaller with accompanying increase in surface energy and subsequent increase in surface reactivity, which leads to thermodynamic instability and ultimately increased toxicity (Rahi et al, 2014). Shape may also affect the toxicity of ENMs. For example, silver nanoplates were reported to have a higher toxicity level in comparison to nanospheres or nanowires (George et al, 2012). The toxicity of the nanoplates was attributed to a high level of crystal defects (stacking faults and point defects) on the surfaces. When cysteine was used as a surface coating, a reduction in toxicity was demonstrated (George et al, 2012). In addition to size and shape, surface characteristics of ENMs enable them to interact differently with biological and environmental surroundings (compared to the bulk-sized particles), which would influence the uptake thereof. Subsequently, surface area, surface chemical composition and surface activity play a decisive role in addition to their size in the hazard identification of ENMs (Yokel & MacPhail, 2011).

The environment in which the ENMs exist may also change their physicochemical properties and, hence, influence their toxicity. These changes may include alterations of the surface properties of the ENMs (e.g., charge), which can result in their aggregation (Handy, 2008; Jiang et al, 2009). For example, it has been shown that TiO_2 becomes hydrophilic after contact with water and forms aggregates in solution due to the coating layer (Auffan et al, 2010). ENPs may also interact with proteins (Saptarshi et al, 2013) or with natural organic matter in the environment and undergo transformations via biotic (e.g., macromolecule degradation) and abiotic (e.g., changes in surface charge) interactions (Unrine et al, 2012), which may, again, change their toxicity.

These and other considerations have made it difficult to assess ENM impact on human and environmental health (Colvin 2003; Maynard, 2006; Maynard et al, 2006; Maynard, 2008; Helmus, 2007; Alkilany & Murphy, 2010) and therefore make it difficult to assess their risk especially in relation to the prediction of long-term effects on human health and on the environment.

Overview of exposure to ENMs

The major source of exposure to, as well as entry routes for, ENMs (e.g., lung, gut and possibly skin) and putative targets (e.g., lung, liver, heart and brain) have been identified (Elsaesser & Howard, 2012). The study of the respiratory and dermal exposure to nanomaterials is important since human skin and lungs (even the gastro-intestinal tract) are constantly in contact with the environment, which increases the exposure time to any possible toxic agents. While the skin is generally an effective barrier to foreign substances, the lungs and gastro-intestinal tract are more vulnerable. Injections and implants are other possible routes of exposure for intentionally administered ENMs (Nuñez-Anita et al, 2014). There are growing concerns regarding exposure to ENMs, not only in the work place during their synthesis and applications, but also in the environment during their disposal (Figure 2.1).

Figure 2.1: Exposure of ENMs that may be toxic to human health and the environment (adapted from Elsaesser & Howard, 2012).

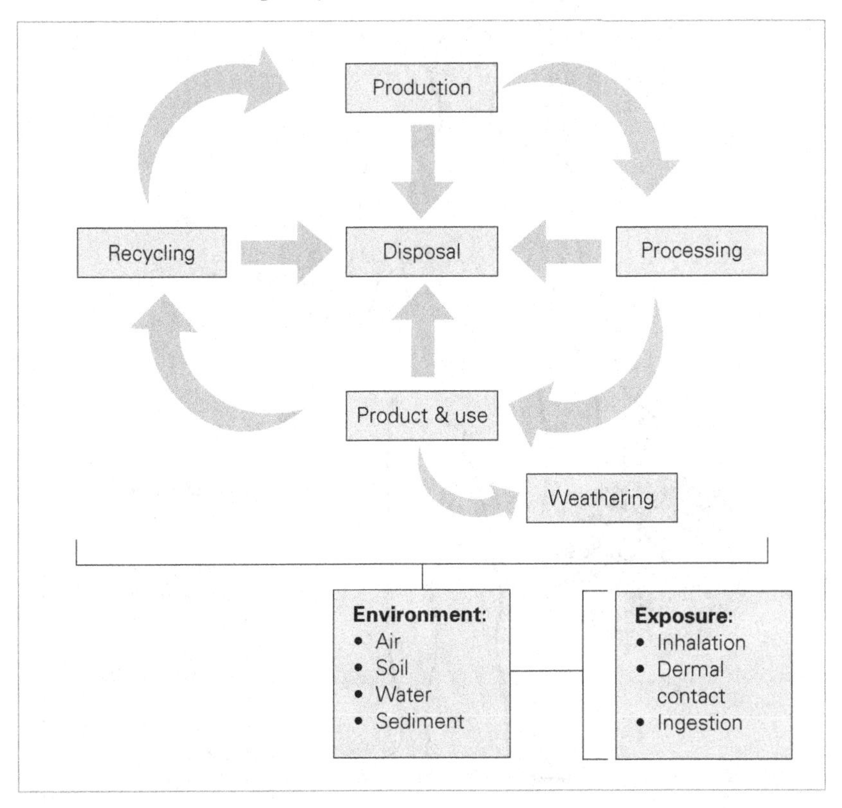

The main concern with ENMs, compared to their larger counterparts, involves their ability to translocate from the site of deposition to other organs, such as brain, blood, liver, spleen and kidneys. ENMs that are not excreted and/or degraded due to their biodurability, may then be biopersistent and, hence, accumulate in different target organs. Therefore, it is important to consider their toxicokinetics and toxicodynamics once inside the human body in order to assess their short-term toxicity, as well as their long-term pathogenicity (Figure 2.2).

Figure 2.2: Toxicokinetics and toxicodynamics of ENMs (adapted from Buzea et al, 2007; Kumar & Dhawan, 2013)

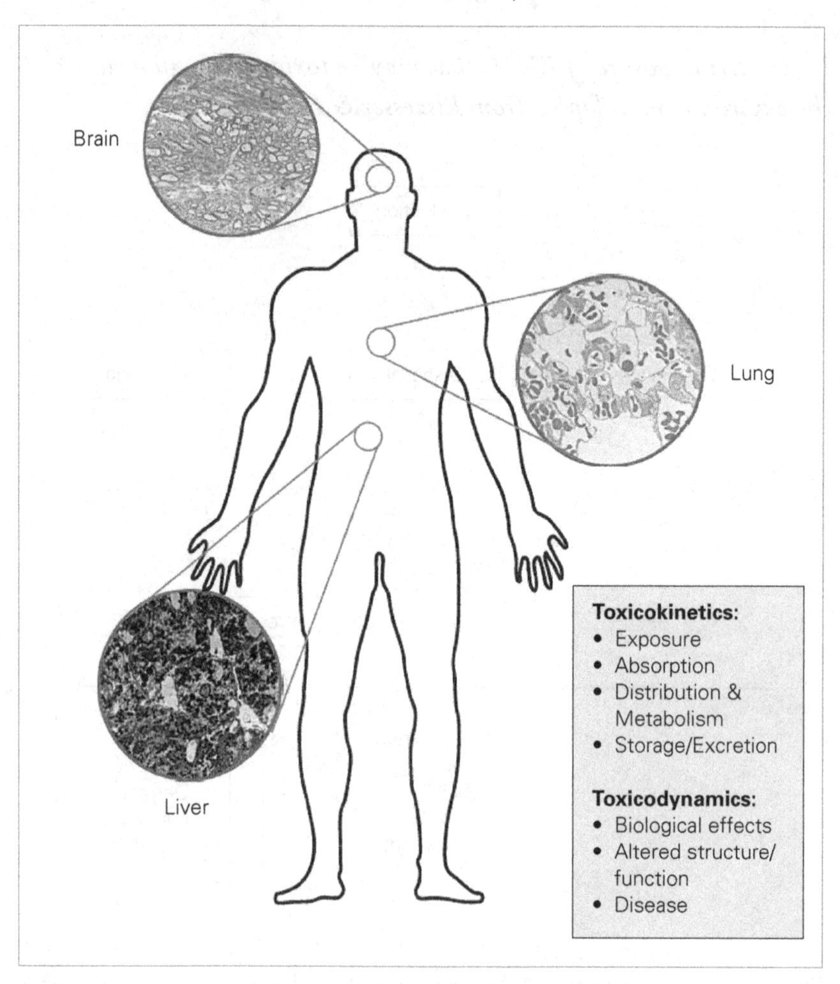

Presently, the most commonly encountered nanomaterials in the work place and/or the environment include TiO_2 (Hussain et al, 2010), cerium dioxide (Park et al, 2008), silicon dioxide (Napierska et al, 2010), zinc oxide (ZnO) (Osmond & McCall, 2010), silver (Ahamed et al, 2010) and carbon nanotubes (CNTs) (Maynard, 2006).

Toxicokinetics

Toxicokinetics in general describes the absorption, distribution, metabolism and excretion (ADME) of chemical toxicants. Due to the ability of ENMs to translocate, toxicokinetics illustrates the translocation through the organism from the route of entry, as well as distribution and accumulation in different organs and, finally, their clearance and excretion. The translocation and the accumulation of ENMs depend, once again, on their physicochemical properties such as type, shape, surface charge and surface coating. This is in addition to the route of exposure, dose and the exposure period (Rim et al, 2013).

Toxicokinetics following respiratory tract exposure via inhalation

The mechanism related to exposure through inhalation revolves around the respiratory system, primarily the nose and throat, which has been identified as a site for quick deposition and translocation of ENMs. Smaller particles are deposited in the upper airways and may be expelled by finger-like cilia, as well as the mucous lining of the trachea and bronchial tubes, which together move particles back up into the throat and nose, from where they are expelled. The smallest ENMs penetrate deeper into the alveolar region and deposit on the epithelium. It has been suggested by Warheit and colleagues that ultra-fine ENMs with a diameter of 50–100 nm may deposit as aggregates due to Van der Waal's forces (Oberdörster, 2010), with a diameter of 200–500 nm, rather than as discrete particles (Goldman & Coussens, 2005).

Once ENMs are deposited via inhalation, they translocate to extra-pulmonary sites, cross the air-blood tissue barrier and enter the blood circulatory system, through which they reach other tissues and organs or travel directly to the brain via the olfactory bulb (Oberdörster et al, 2005; Jain et al, 2012). Depending on their size and other physicochemical properties, these ENMs translocating via the latter route will then be able to travel along axons and dendrites to gain direct access to the central nervous system. Neuronal uptake of inhaled ENMs may, therefore, take place via the olfactory nerves but also via the blood-brain barrier (Oberdörster et al, 2005; Elder et al, 2006; Buzea et al, 2007). The effect of size on the bio-distribution of gold NPs (AuNPs) was investigated via an inhalation study using Sprague-Dawley rats. Although no toxic response was observed, the

AuNPs did translocate from the lungs, and smaller sized AuNPs were detected in the liver, spleen, brain, testes, blood and brain (Han et al, 2015).

The small particle size, with the corresponding high surface areas, may not be the only factor responsible for triggering a toxic response or the translocation ability of ENMs from the lung to other organs. Ma-Hock and colleagues investigated the effect of a glutaraldehyde coating on the inhalation toxicity and translocation of $CdS/Cd(OH)_2$ core shell quantum dots (Ma-Hock et al, 2014). Quantum dots are NPs that consist of a crystalline core surrounded by a cap. Rats were exposed in a head-nose-only chamber. The coated quantum dots were administered at the maximum concentration that was technically feasible. The exposure was for 6h/d, over five consecutive days, followed by a three-week exposure-free period. Organs and excretions were examined to determine the Cd content after the exposure. The Cd was mostly present in the lungs and faeces. Even though the quantum dots were small enough to facilitate crossing of biological barriers, translocation of particles from the respiratory tract was not observed when using an Inductively Coupled Plasma Mass Spectrometer (ICP-MS). The broncho-alveolar fluid showed minimal inflammation. Microscopy of the larynx indicated minimal to mild epithelial alteration. The quantum dots did not elicit significant effects other than the toxicity of the Cd^{2+} ion itself.

Toxicity during the development of the immune system is significant due to the influence it has on disease contracted later in life. There are concerns that prenatal exposure to nanomaterials may have adverse effects on the development of the immune system of infants. Consequently, El-Sayed and colleagues (El-Sayed et al, 2015) exposed pregnant Institue of Cancer (ICR) mice to carbon black nanoparticles (CB-NP), via intranasal instillation. This treatment was performed during the development of the embryonic thymus and spleen, i.e. on the 9th and 15th gestational days (GDs). Therefore, maternal respiratory exposure to CB-NP was assessed during critical periods of offspring development. The thymus (primary lymphoid organ) and spleen (lymphoid organ involved in the initiation of the immune response) were collected from the offspring on the 1st, 3rd and 5th postnatal days

(PNDs). The results indicate dysregulation of lymphocyte populations and alteration of lymphocytic phenotypes. Gene expression analyses confirmed a significant effect of CB-NP on the *Traf6* gene, which is an adapter molecule that mediates signals and is required for the induction and maintenance of peripheral T-cell self-tolerance. It was proposed that the observed immune-stimulation could be mediated by allergic reactions and inflammatory responses during the early life of the newborn offspring. It was also noted that the NP dose level of the maternal exposure, in addition to the route of exposure, as well as the stage of gestation during which exposure occurred, appeared to promote the offspring's immune response to the allergen.

Toxicokinetics following gastro-intestinal exposure via ingestion

The uptake of ENMs via the gut is possible (Florence, 2005; Martirosyan & Schneider, 2014) and seems to be size dependent (Jani et al, 1992). The gastro-intestinal (GI) tract is a complex barrier-exchange system, where the epithelium of both the small and large intestines is in close proximity to ingested material that is absorbed by the villi. The extent of ENM absorption in the GI tract is affected by size, surface chemistry, charge, length of administration and dose (Buzea et al, 2007). Exogenous sources of ENMs in the GI tract include particles from food (e.g., titanium oxide used as a colourant), pharmaceuticals, water or cosmetics (e.g., toothpaste and lipstick), dental prosthesis debris (Ballestri et al, 2001) and inhaled particles (Takenaka et al, 2001).

The duration of exposure of orally administered ENMs could also influence their translocation and accumulation. For example, Sprague-Dawley rats orally exposed to 10 nm and 25 nm citrate-stabilised silver nanoparticles (AgNPs) daily for four weeks, followed by a four-month recovery showed that silver content in the blood, liver, ovaries, kidneys and spleen decreased over time. However, silver in the brain and testes did not clear well (Lee et al, 2013). Similar accumulation in the brain and testes after an eight-week recovery period was observed in rats treated daily for 28 days via oral exposure of <15 nm PVP-coated and <20 nm non-coated AgNPs (Van der Zande et al, 2012). On the other hand, a single dose of orally administered 8 nm citrate-capped

AgNPs in Sprague-Dawley rats was primarily excreted in the faeces and not absorbed via the gastrointestinal tract (Park, 2013). Similar observations were made on the accumulation of orally administered ZnO nanoparticles. Sprague-Dawley rats that received ZnO orally by a single dose exhibited an increase in plasma concentration, in a dose-dependent manner, which returned to normal within 24 hours (Chung et al, 2013) while rats receiving a 90-day repeated oral dose did not return to their normal ZnO plasma concentrations. This suggests that these nanoparticles could accumulate. These results also imply that a repeated dose of orally administered ENMs increases the likelihood of translocation and accumulation.

In a study on the effects of size and surface charge on the pharmacokinetics of ZnO nanoparticles, negatively charged ZnO showed higher absorption than positively charged ZnO nanoparticles, while differences in size did not affect pharmacokinetic profiles or biodistribution. However, particle size did influence excretion, where 20 nm nanoparticles were eliminated more rapidly than the 70 nm nanoparticles studied (Paek et al, 2013).

The translocation of AuNPs in daphnids following exposure through the GI-tract was also investigated and it was concluded that although the AuNPs were retained in the gut lumen, internalisation into gut epithelial cells was not observed (Khan et al, 2014).

Toxicokinetics following dermal exposure via skin contact

The skin is a structured organ that consists of layers including the epidermis, dermis and subcutaneous layer. The epidermis has a scaly appearance with pores for sweat, sebaceous glands and hair follicle sites. The mechanism for dermal uptake of ENMs includes crossing of skin pores and the first layer of protection (stratum corneum, SC), as well as localisation around the root of the hair follicle or broken skin. The 10 µm thick and keratinised layer of dead cells above the epidermis that is the stratum corneum may be the rate-limiting barrier to ENMs. Once the ENMs have lodged inside the avascular epidermis, they are not removed by phagocytosis (Monteiro-Riviere & Inman, 2006; Jain et al, 2012). If ENMs are not removed via phagocytosis, they could ultimately enter the systemic circulation and the lymphatic system (Oberdörster et al, 2005).

The ENMs that seem to be able to penetrate the dermis (Rouse et al, 2007; Ryman-Rasmussen et al, 2006) include fullerenes (Xia et al, 2010) and quantum dots (Stern & McNeil, 2008), which appears to be dependent on the size and surface coatings (Ryman-Rasmussen et al, 2007). Coated and uncoated zinc oxide (ZnO) NPs were assessed to determine the extent of their penetration into the epidermis of human volunteers (Leite-Silva et al, 2013). Although limited penetration was observed, it was not sufficient to affect the redox state of the cells. The effects of surface coating on the penetration and translocation of TiO_2 through the skin was also investigated since commercially available cosmetics and sunscreens often contain TiO_2 that may gain access through hair follicles or wounds and lesions (Crosera et al, 2009). Sunscreens, which contained 8% TiO_2 (10–15 nm), onto the skin of humans did not penetrate (Tan et al, 1996; Tsuji et al, 2006). In contrast, those containing oil-in-water emulsions did penetrate, where higher penetration was observed for hairy skin at the hair follicle site or pores (Tsuji et al, 2006).

The SC of human skin was also reported to be treated with two types of silica samples (Iannuccelli et al, 2014). This treatment was performed in order to determine the impact of the silica NP surface polarity, as well as the effect of human hair follicle density, in relation to the extent of skin penetration. The bare hydrophilic silica (B-silica) was smaller in size than the lipid-coated silica (LC-silica) with the surface characteristics that were modified from hydrophilic to hydrophobic. The silica was applied to two different skin regions with different hair follicle densities, i.e., the volar and dorsal (hairy) side of the forearms of volunteers. Nano-sized structures were located in the broad spaces between corneocytes. Only 10% of the B-silica permeated the deepest SC layers, indicating that silica was retained in the upper SC layers, regardless of the hair follicle density. LC-silica penetrated to a greater extent into the deeper SC layers, i.e., 42% for the volar and 18% for the dorsal application. It is generally accepted that the smaller the NP, the greater the penetration of the skin layers. However, it was concluded that the NP surface polarity played a greater role than the size of the NP with regards to the route, as well as the extent of penetration.

In another study, the influence of nano-TiO_2 on trans-dermal drug delivery was assessed (Peira et al, 2014). The investigation made use of the effect of naked or coated nano-TiO_2 on the flux of the antifungal agent Amphotericin B (AmB) through porcine skin. Nano-TiO_2 may penetrate the first layer of the skin (SC) and generate free radicals under low UV radiation. Coated nano-TiO_2 would cause significant differences in skin permeation due to the alterations of the physio-chemical features, which would influence the biological activity. Consequently, it was observed that naked nano-TiO_2 enhanced the drug flux through the skin. The naked sample had a positive surface charge and was found to adhere irreversibly to the negatively charged corneocytes of the SC. The silane/silica-coated samples were negatively charged and were easily removed by washing. It was concluded that the enhancer effect of nano-TiO_2 could be turned up or down, in order to favour or impair drug penetration. This was achieved by modulating the particle net surface charge (e.g. coating), as well as its oxidative potential (e.g. crystalline phase). In contrast to drug delivery, translocation of bioactive molecules was not recommended for applications including cosmetics and sunscreen lotions.

Toxicokinetics following exposure via intravenous injection

The primary route of entry also appears to influence the toxicokinetics of the nanoparticles. For example, a study on the biodistribution of titanium dioxide nanoparticles (TiO_2 NPs) in Wistar rats by single dose and repeated dose (on five consecutive days, via oral exposure) showed minimal absorption in the gastrointestinal tract. However, when the animals were treated using tail vein intravenous injection, the ENMs circulated to all tissues evaluated and, in particular, to the liver, spleen and lungs. After a 90-day recovery period, slow tissue elimination of the TiO_2 NPs was observed, as well as redistribution from the liver to the spleen (Geraets et al, 2014).

The shape of ENMs will also influence their translocation, biodistribution and accumulation following intravenous injection. For example, it has been reported that silica ENMs of 70 nm diameter, with similar chemical composition and surface charge, but of diverse shapes, were retained differently after intravenous injection *in vivo* (Huang et al, 2011). Short-rod ENMs were found to be trapped in the liver, while long-rod ENMs were found in the spleen and remained longer in the blood, before being excreted through urine/faeces at a slower pace.

Toxicodynamics

Toxicodynamics describes the adverse effects that a toxicant has on an organism. Once again, this will be determined by the physicochemical properties of ENMs such as size, shape and surface reactivity. Such effects are studied using both *in vitro* cell models in relevant immortal and primary cells, as well as *in vivo* models, generally in mice and rats. The choice of *in vitro* cell lines is often based on the identification of target organs in toxicokinetic studies. Once again, an increasing number of scientific reports have highlighted the need to understand the interactions between different types of ENMs and cells in relation to their physicochemical properties such as size, shape, aspect ratio, surface chemistry and activity of the nanomaterial (Lewinski et al, 2008).

Oxidative stress, mitochondrial and DNA damage in vitro

The generation of excessive reactive oxygen species (ROS) and oxidative stress is the most frequently reported mechanism for nanoparticle toxicity (Manke et al, 2013). ROS is the term used for O_2-derived free radicals and the mitochondria are the major source of intracellular ROS. Excess ROS production initially induces an antioxidant response, but at intermediate levels it will induce a pro-inflammatory response. At higher levels of oxidative stress, apoptosis results from mitochondrial damage and the release of pro-apoptotic factors. In addition, other forms of injury may result, including irreversible oxidative damage to proteins, lipids and DNA.

In vitro studies have identified oxidative stress as the mechanism of toxicity for various types of nanoparticles and in various cell types. Zinc oxide nanoparticles were shown to induce ROS in a dose- and time-dependent manner in normal skin cells, which lead to abnormal autophagy of vacuoles, mitochondrial damage and dysfunction, and finally cell death (Yu et al, 2013). Human lung carcinoma cells treated with nickel oxide nanoparticles showed an increase in ROS and lipid peroxidation (Horie et al, 2011). Recently, human amnion epithelial WISH (Wistar Institute, Susan Hayflick) cells were exposed to TiO_2, which resulted in a reduction in cell viability, as well as morphological alterations, in addition to a compromised antioxidant system, intracellular ROS production and DNA damage (Saquib et al, 2012).

The ability of an agent to cause DNA damage, known as genotoxicity, can occur either indirectly via oxidative stress or directly if the particles are small enough to pass through cellular membranes and gain access to the nucleus, or during mitosis when the nuclear membrane breaks down (Singh et al, 2009). *In vitro* oxidative DNA damage has been observed in cells exposed to gold nanoparticles (Di Bucchianico et al, 2014), copper oxide nanoparticles (Alarifi et al, 2013), as well as cerium oxide nanoparticles (De Marzi et al, 2013; Mittal & Pandey, 2014).

Oxidative stress, mitochondrial and DNA damage in vivo

A few studies have also been conducted to investigate whether nanoparticles can induce oxidative stress *in vivo*. Instillation of nickel oxide nanoparticles into the trachea of rats, followed by collection of bronchoalveolar lavage fluid displayed an increase in lipid peroxidation and lactate dehydrogenase leakage (Horie et al, 2011). Long-term exposure of mice to inhaled nickel hydroxide ENMs induced an up-regulation of antioxidant and inflammatory cytokine genes in both pulmonary and extrapulmonary organs. It also increased mitochondrial DNA damage in the aorta and inflammation in bronchial lavage fluid (Kang et al, 2011). This study showed that the induced oxidative stress was not limited to the lung but was also found in the cardiovascular system, and could ultimately result in the development of atherosclerosis in the mice.

Bare silver nanoparticles were shown to induce mitochondrial membrane damage and, in addition, bare as well as polyvinyl/pyrrolidone-coated silver nanoparticles induced oxidative DNA damage in the nematode *Caenorhabditis elegans*, which is used as a model of environmental toxicology (Ahn et al, 2014). In a study that compared the toxicities of titanium dioxide NPs, zinc oxide NPs, and silicon dioxide NPs to *C. elegans*, all three nanoparticles resulted in ROS production to various degrees (Wu et al, 2013).

Since the complexity increases when moving from *in vitro* to *in vivo* models, ENP toxicodynamics *in vivo* is still at an early stage. Recent *in vivo* studies have assessed exposure of AgNPs in Sprague-Dawley rats, where the genotoxic potential was shown after an inhalation exposure for 90 days (Kim et al, 2011). The DNA damage to lung cells, after repeated inhalation of AgNPs, was also shown over a 12-week exposure in a whole-body inhalation chamber (Cho et al, 2013). It was noted that the level of DNA damage was significantly higher for the high-dose group, as determined by the Comet assay.

The respirable single-walled or multi-walled carbon nanotubes were also studied using electron spin resonance (ESR) in order to assess whether or not exposure would cause formation of free radicals in the lungs, heart and liver *in vivo*. It was found that exposure did indeed trigger interactions of early inflammatory responses, generation of free radicals and oxidative stress (Shvedova et al, 2014a). Acute exposure to SWCNTs was also reported to accelerate the formation and growth of lung carcinoma of mice, i.e., those treated for 48 hours prior to receiving Lewis lung carcinoma cells (Shvedova et al, 2013). Caution was also advised regarding MWCNT inhalation exposure, based on analyses of lung tumour formation in B6C3F1 mice (Sargent et al, 2014). The long-term effects of ENMs containing carbon were also assessed in mice over a one-year period, with post-exposure comparisons (Shvedova et al, 2014b).

Additional recent *in vivo* studies, including the occupational exposure to nanosilica, was monitored in the workplace, although no cytotoxic or genotoxic analyses were performed (Kim et al, 2014). Other nanometal oxides from workplace surfaces in a semiconductor research and development facility were also analysed to determine accessibility with regards to cutaneous exposure (Brenner & Neu-Baker, 2015). However, again, no cytotoxic or genotoxic analyses were performed.

IMMUNOTOXICITY

The immunotoxic effects of a number of ENMs were also investigated. For example, intratracheal instillation of TiO_2 produced no change in the levels of IFN-γ and IL-4 cytokines, but produced increased macrophage accumulation, extensive disruption of alveolar septa and slight alveolar thickness, as well as expansion hypermia. In addition, up-regulation of the GATA-3 and down regulation of the T-bet transcription factors, the latter of which is linked to Th1 differentiation and the former of which is a key regulator of Th2 development. Th1 cells produce IFN-γ and Th2 cells produce IL-4. It was, therefore, proposed that the Th1/Th2 cytokine imbalance may be a mechanism associated with the TiO_2-induced immunotoxicity of the respiratory system (Chang et al, 2014). A similar imbalance of Th1 and Th2 cell development was also observed in mice following exposure via oropharyngeal aspiration with NiNPs and multi-walled carbon nanotubes (MWCNTs) (Glista-Baker et al, 2014). A 28-day repeat-dose intravenous AgNP administration in rats showed a reduction in keyhole limpet hemocyanin (KLH)-specific IgG, suggesting a suppression of the immune system (Vandebriel et al, 2014).

In addition, the effects of exposure to ENMs on pre-existing respiratory diseases was also investigated and was found that the combination of exposure to TiO_2 and respiratory allergens in pre-sensitised rats could potentiate neutrophil inflammation and lymphocyte responses. In contrast, the eosinophilic responses were suppressed in the airways. Since the responses were strain dependent, it was concluded that genetic factors influenced the susceptibility to the particular type of inflammatory responses. It was, therefore, concluded that the genetic variation appears to compound the problems associated with health risk assessment related to ENMs exposure (Gustafsson et al, 2014)

EFFORTS ON HUMAN AND ENVIRONMENTAL HEALTH IN SOUTH AFRICA: THE EXTENT OF NANOTECHNOLOGY ACTIVITIES IN SOUTH AFRICA

Like many other countries in the world, South Africa has increased its investment in nanotechnology and nanoscience research and training by establishing centres for nanotechnology and funding nanoscience and nanotechnology in higher-education institutions. For example, the South African government, through the Department of Science and Technology (DST), has established a few centres for nanotechnology, which serve as strategic partners of research in academia and development in industry, aimed at addressing national priorities highlighted by the national nanotechnology strategy (Nyokong & Limson, 2013). The ten major centres for nanoscience and nanotechnology research and development are: the universities of Cape Town, the Western Cape, Stellenbosch, the Witwatersrand, Pretoria, Zululand and Limpopo; the other centres are the iThemba Laboratories, the Council for Scientific and Industrial Research (CSIR), and the Council for Mineral Technology (MINTEK) (Campbell, 2006, CSIR Communication, 2007 and MINTEK, 2014). These centres are national facilities expected to coordinate and promote nanotechnology across the country and their focus includes the development of research platforms, promoting collaborative networks, addressing human capital development and bridging the innovation gap.

In line with the rapid progress in nanotechnology research and development, more and more universities in South Africa have also begun offering degree programmes in nanotechnology, through a government-led initiative called the National Nanoscience Postgraduate Teaching and Training Programme (DST Communication, 2013). Through this programme, which is also funded by DST, four South African universities (the University of Johannesburg, Nelson Mandela Metropolitan University, the University of the Free State and the University of the Western Cape) offer a registered MSc nanoscience degree with specialisation in nanochemistry, nanophysics or nanobiomedical science (Nyokong & Limson, 2013). It is expected that graduates of this programme will provide a pool of nanonscientists with the knowledge and technical and scientific skills that are lacking in existing innovation centres and research groups since training in South Africa is on specific research projects only.

Table 2.1: Chronology of nanoscience and nanotechnology in South Africa (modified from Gulumian, 2009).

Year	Activity
2002	The South African Nanotechnology Initiative (SANi) was formed.
2003	South Africa's Advanced Manufacturing Technology Strategy (AMTS) was launched.
2005	The Department of Science and Technology published the National Nanotechnology Strategy.
2007	Nanotechnology innovation centres were established.
2009	National Nanoscience Postgraduate Teaching and Training Programme was launched.
2014	UNESCO-UNISA Africa Chair in Nanoscience and nanotechnology was launched (UNISA, 2014).

It is also expected that as progress for nanotechnology research and development accelerates, more and more universities in South Africa will begin to offer degree programmes in nanotechnology, and curricula will evolve to keep abreast of emerging areas. Table 2.1 is a chronology of activities arising as a result of strong interest in nanoscience and nanotechnology in South Africa (Gulumian, 2009).

The National Research Foundation (NRF) of South Africa has also launched the National Nanotechnology Equipment Programme (NNEP) in support of the National Nanotechnology Strategy, by supporting the acquisition of research equipment for the analysis and characterisation of nanomaterials, as well as promoting regional, national and international research collaborations in nanoscience (NRF, 2014). In 2014 another major collaboration was launched, the UNESCO Africa Chair in Nanoscience and Nanotechnology, which is a trilateral partnership between UNESCO, the University of South Africa and iThemba Labs (UNISA, 2014). This is a group of multi-disciplinary research laboratories administered by the NRF that serves as a think tank and a bridge over the existing knowledge crevasse between academia and civil society, local communities and policy-makers.

The rapid rate of production, and the manufacturing of and research into nanomaterials is expected to generate large volumes of nanowaste, requiring appropriate processing, recycling and disposal methods (Allan et al, 2009; Simonet & Valcárcel, 2009). Consequently, the DST has emphasised the need for an assessment of the risk associated with nanomaterials being synthesised in the country presently. Such an assessment is indeed critical for addressing the potential, unintended consequences of nanotechnology.

Unfortunately, risk assessment-based research was initially overlooked South Africa in favour of fundamental investigations and the application of nanoscience and nanotechnology (Claassens & Motuku 2006). What this means, therefore, is that like the rest of the world South Africa is yet to develop a national research strategy to investigate nanotechnology risks (Hansen 2010; Gulumian et al, 2012). Fortunately for South Africa, there is no large-scale production of ENMs yet, but only laboratory-based synthesis, which is small (Wild, 2013; de Jager et al, 2014). Moreover, current efforts to address health and safety concerns in nanotechnology development and in nanotoxicology of ENMs are gradually picking up speed, in line with the speed of nanotechnological developments in the country (CSIR Communication, 2009a; Wild, 2013). For instance, the National Institute for Occupational Health (NIOH), CSIR and a few South African universities are engaged in research and training to generate data for safety evaluation, toxicologic evaluation of potential effect on human health, risk assessment to support risk-management decision-making, and to address inadequacies in existing regulations (CSIR Communication 2009b; Gulumian, 2009).

The quest for fully developed nanotechnology in South Africa that is internationally competitive has gained momentum with the establishment of programmes such as the South Africa Nanoscience and Nanotechnology Summer School and Nanotechnology Public Engagement Programme (NPEP) (SAASTA, 2001; SAASTA, 2014). All these programmes are led or at least partly sponsored by the DST, as part of the implementation of the National Nanotechnology Strategy. The summer school participants (postgraduate students and experienced scientists) have the opportunity to train on state-of-the-art instruments and to hold discussions and debates around issues related to a specific theme. Given consumers' increasing exposure to the flood of nanotechnology products entering the South African market there is a clear need for the public to be informed in order to be able to embrace the promise of nanotechnology while at the same time being fully aware of its potential to cause costly long-term problems to human health and the environment. The NPEP was established for this reason, and serves as a communication programme, the purpose of which is to inform, educate and engage the public on nanotechnology and its potential societal impact (SAASTA, 2001).

NANOTECHNOLOGY INDUSTRY/COMPANIES

South Africa is among the first countries globally to have an official nanotechnology strategy (DST Communication, 2008). This nanotechnology strategy and its associated 10-year plan branches into two main categories: the social branch, which deals with health, water and energy research, and the industrial branch, which investigates mining and minerals, chemical and bioprocessing, and the development of advanced materials. The social branch is aimed at poverty alleviation, rural development, and health and sanitation, whereas the industrial branch is concerned with the generation and utilisation of clean, low-cost energy. Two National Nanotechnology Innovation Centres based at CSIR and MINTEK have been established specifically for this purpose (CSIR Communication, 2007; MINTEK, 2014). However, most of the work being done at these sites is still in the area of nanoscience or research only, and this is only expected to change as more applications are developed.

The practical application of nanoscience in South Africa has seen the birth of a few companies and departments within companies conforming to any of the two branches described above (Campbell, 2006). For example, companies such as Sasol, which is interested in more efficient hydrogen fuel-cells and Element Six (formerly De Beers Industrial Diamond Division) have established nanotechnology research divisions. Statutory bodies such as the Water Research Commission and the Medical Research Council also conduct nanotechnology research on products suitable for water purification and for the manufacturing of diagnostics and therapeutics, respectively (Campbell, 2006; NPEP, 2014). A Google web search shows that a few private nanotechnology companies also exist within South Africa and some of them are briefly described below:

Comar Chemicals manufactures iron-based nanoparticle colloids and carboxylates. These products are used as synthetic rubber catalysts, paint driers and polyester accelerators and other proprietary applications (NanoBugle, 2012; Comar Chemicals, 2014).

Nanotech Water Solutions is a water purification company, which delivers water purification, water treatment, and disinfection and filtration solutions to the southern African market (NanoBugle, 2012; NanoTech Water Solutions, 2014).

Plascon and **Dulux,** paint manufacturers, are using nanotechnology to improve paint properties (NPEP, 2014).

Nanotech Energy South Africa provides affordable energy solutions using renewable energy sources (Nanotech Energy, 2014).

Rubber Nano Products (Pty) Ltd manufactures specific activators for rubber vulcanisation that allow zinc reduction while achieving significant industrial advantages during rubber manufacture (Rubber Nano Products, 2014).

Nano Bubble Technology is a company that exploits new nanotechnology to help reduce greenhouse gases, harmful emission and pollutants using cheap environmentally safer fuel (Nano Bubble Technology 2014).

HEALTH RISK ASSESSMENT AND MANAGEMENT INITIATIVES

As indicated above, South Africa through its innovation policy framework is engaged in significant nanotechnology research initiatives and has established national activities aimed at promoting social and economic growth. Consequently, these new technological developments have inevitably led to the generation of a new waste stream, of waste-containing nanomaterials, herein called nanowaste.

So far, risk assessment of ENMs in South Africa has focused mainly on emissions during production and use, while nanowaste management has received little, if any, attention. Currently, major nanowaste generators in the country are research and manufacturing facilities (De Jager et al, 2014). Nanomaterials synthesised at these sites are often likely to contain other chemicals such as metals or organic chemicals used as catalysts or for doping (Chen & Wang, 2013). Such nanowaste contains both nanomaterials and associated hazardous chemicals in varying concentrations and likely requires extensive characterisation (Allan, 2013). Unfortunately, there is no general international consensus on how to determine whether a particular nanomaterial is classified as hazardous or not. As a result, the legislation relating to current waste disposal practices has not been altered to take into account nanowaste storage, containment and disposal (Allan et al, 2009).

A report presented to the Department of Environmental Affairs and Tourism indicates that the rapid generation of nanowaste in South Africa has already started to challenge established waste management practices and technologies with respect to their suitability to nanotechnology (Oelofse & Musee, 2008; Musee, 2011). Existing waste regulations do not contain any specific reference to ENMs, and risk managers are uncertain about both the composition of nanomaterials and the various possibilities for modifying nano-objects. Both these factors can affect ecosystems and the environment in unpredictable ways. The global lack of scientific data and knowledge concerning risk assessment of ENMs in different environments has made it difficult to develop legislation and policies that can address the disposal of new waste streams (nanowaste in this case), or modifications to current waste management systems (Aschberger et al, 2011).

A major challenge in dealing with nanowaste is the fact that several derivatives of the same material can be produced using different manufacturing processes. In this way, there are differences in the potential risks posed by various nanomaterials and nanowaste generated exhibit a range of variable toxicological and ecotoxicity characteristics (Ju-Nam & Lead, 2008). The characteristics of ENMs such as transportability, persistence, aggregation and disaggregation have already shown potential for bioaccumulation, toxicity and mutagenicity (Bystrzejewska-Piotrowska et al, 2009). As Boldrin and colleagues indicate, ENMs may be released into the environment during all stages of regular waste management such as collection, recycling, incineration and land filling. Consequently, nanowaste may require a waste management approach that differs from the conventional large-scale treatment of wastes (Boldrin et al, 2014). In particular, in the recycling of nanowaste ENMs may be released and emitted during shredding, milling, sorting and thermal processing, resulting in contamination of the working environment (Köhler et al, 2008). Furthermore, most poor communities in South African cities, just like in other developing countries, derive some livelihood from salvaging solid waste for recycling. These communities could be directly exposed to ENMs from sorting solid waste and from the consumption of nanoparticle-contaminated food waste (McLean, 2000; Wilson et al, 2006; Medina, 2010; Schenck & Blaauw, 2011).

The latest Earth Engineering Centre of Columbia University (EEC) study report describes the many ways in which waste can be used as an energy resource that could even assist in the reduction of greenhouse gas emissions (Themelis et al, 2011; Vanderveen, 2014). Despite the fact that this information is now in the public domain, corporate transparency is still an obstacle with regard to nanowaste management (Allan et al, 2009; Bystrzejewska-Piotrowska et al, 2009). Without knowing how companies plan to use and store recycled and non-recycled nanomaterials, it is difficult to develop regulations or modifications relating to current waste management systems.

In a recent survey of nanotechnology activities within South Africa conducted by the NIOH on behalf of the DST (De Jager et al, 2014), it was found that currently all users of ENMs (industry, academia and domestic) still rely on conventional ways of waste disposal. For example, either a private commercial company is used to dispose of nanomaterials-related waste or the nanowaste is treated like domestic waste due for disposal by the local municipality. A small percentage of respondents in the same survey reported either recycling/storing or incinerating/autoclaving nanomaterial-related waste as the primary method of waste management. It is, therefore, not unreasonable to think that as the production of and possible applications for nanomaterials increase, waste material related to these activities will increase, and possibly become ubiquitous in the environment as a result of poor or inadequate management. However, international recommendations are to regard nanowaste as hazardous waste and assign those potentially exposed to it a high degree of protection. This entails, as a minimum, nanowaste to be double-bagged, enclosed in a rigid impermeable container and preferably bound within a solid matrix (Edwards et al, 2007; Allan et al, 2009; Kreider et al, 2013).

At present the effects of NP on human health are unclear (SCENIHR, 2009; EFSA Scientific Committee, 2011). However, there are legitimate concerns the world over that the nano-sized particles employed in this new technology will have negative effects on the health of humans, animals and the environment. Of major concern currently is that the wide production and utilisation of ENMs is rapidly overtaking efforts to evaluate their toxicity to humans and the environment. Internationally, there are efforts to develop standardised toxicity testing and risk assessment methods for ENMs. It is therefore anticipated that as scientific knowledge improves, it will be possible to classify nanomaterials into specific risk categories that can be used for category-specific risk assessments.

At present, risk categorisation of nanomaterials is not possible; however, new significant developments such as accelerated research by the global scientific community aimed at acquiring exposure and hazard data on a wide range of manufactured ENMs and nanoproducts have come to light. There is currently also very limited research on safety, health- and environment-related aspects of ENMs in South Africa. There is currently no regulatory guidance on nanowaste management in South Africa; however, South African researchers are assessing the risks of different ENMs on an equal footing with the rest of the industrialised world (Wild, 2013). A few South African initiatives can be cited that are aimed at identifying priorities and supporting effective research that can facilitate risk assessment and risk management decision-making. The First South African National Workshop on Nanotechnology Risk Assessment was held in Pretoria in 2009. The objectives were to:

- establish a national research platform network;
- identify key tasks;
- establish an operational research framework;
- develop evaluation metrics for proposed/agreed goals; and
- raise the level of awareness among a diverse group of stakeholders on the need for evidence-based risk assessment to support nanotechnologies and nanosciences.

Following this workshop, the Nanotechnology Research Platform Network was established by the DST (CSIR Communication, 2009a; b; Gulumian et al, 2012). The DST is working in collaboration with NIOH, CSIR, University of Pretoria and North West University to lead the research efforts into understanding the health, safety and environmental risks associated with nanotechnology in South Africa using a three-pronged strategy that is aimed at:

- development of key components of a risk assessment for nanotechnology and nanosciences research platform in South Africa;
- baseline studies on published scientific studies on nanotoxicology; and
- an inventory of the nanoproducts and nanomaterials in use, production or imported in South Africa.

In addition, DST has recently completed a baseline study on how to acquire information on published scientific studies on nanoscience and nanotechnology in South Africa through NIOH's Toxicology and Biochemistry Research Section (De Jager et al, 2014). An inventory obtained through this survey of the nanoproducts and nanomaterials in use, production or imported in South Africa shows that 82% of the nanomaterials produced is utilised in research, while the remaining 18% is utilised for commercial application. An ongoing project to compile a public inventory of nanotechnology activities and nanotechnology products already on the South African market is also in the pipeline. It is anticipated that once the Nanotechnology Research Platform Network is fully operational, nanotechnology research and risk management in South Africa will be in par with best international practices.

CONCLUSION

The broad spectrum of various ENM-cell interactions affects many different cellular components and functions, e.g., mitochondria, cytoskeleton, membrane currents, ROS production and intracellular calcium. ENMs are able to not only interact passively with cells, but also actively engage and mediate the molecular processes that are essential for regulating cell functions. These changes in cellular activity may be deemed stressful to the cell and, therefore, create a latent toxicity. However, under certain circumstances, these adverse effects are tolerated for a short time due to necessity, e.g., cancer diagnostic imaging or chemotherapy delivery. ENMs may even stimulate some cellular functions, instead of disrupting them, where ENM uptake implies increased activity of the cellular machinery, e.g., an increase in ROS production, an increase in intracellular calcium and, in general, an enhancement of cellular activity. Hence, the balance between detrimental and beneficial effects of ENMs is a critical point in the various applications.

South Africa has embraced the promise of nanoscience and the exciting possibilities of nanotechnology, as evidenced by the amount of activity associated with its national nanotechnology strategy. The promotion of regional, national and international research collaborations in nanoscience will indeed render the country internationally competitive. South Africa is to be commended on the education and training initiatives, and outreach programmes, used to create a skilled workforce and informed public in line with the rapid growth of nanotechnology. The private sector is also showing an increase in terms of its nanotechnology offerings, which enhance the country's global competitiveness. However, nanotechnology is not without its fair share of problems, emanating from the potentially detrimental effects of the nano-sized particles employed in this new technology. From the discussion presented above, it is obvious that the burgeoning South African nanotechnology industry must work hand in hand with risk management specialists. This means that efforts to establish occupational exposure limits for ENMs and proposal of appropriate nanowaste management must be intensified in order to catch up with the rapid growth of nanotechnology. Currently, little is known about appropriate ways of cleaning up nanomaterial spills and disposing of nanowaste in a research and development environment, not to mention on an industrial scale. In the absence of occupational exposure limits, at least a precautionary risk management approach should be set in place in order to safeguard both the environment and the health of researchers and workers handling nanomaterials. In conclusion, South Africa needs to prioritise its Nanotechnology Research Platform Network and fast-track identification of knowledge gaps in qualitative nanomaterial risk assessment in order to be able to draw up a strategy on nanomaterial risk management.

FUTURE CONSIDERATIONS

This chapter has summarised how ENMs can potentially interact or interfere with biological processes inside the body. However, one still needs to take into account the potential interactions that may occur due to non-degradable or slowly degradable ENMs accumulating in the body and/or organs. In addition, another issue to consider is the life-cycle stage at which these events may occur, whether it be during the anticipated use, disposal or extraction stage. Ultimately, the nano-research community must come to a consensus regarding the best way to measure the toxicity ENMs in studies and risk analyses.

RECOMMENDATIONS

There is a need to understand the interactions between different types of ENMs and cells, in relation to their physicochemical properties, e.g., size, shape, aspect ratio, surface chemistry and activity of the nanomaterial. However, *in vivo* toxicodynamic studies of ENMs are still at an early stage in development. Therefore, more research must be done, e.g., on methods of administration, means of uptake, as well as the body's clearance mechanisms. If one were to compare what is known about natural nanomaterials and apply that information to modified/synthetic ENMs, it might be possible to predict the functionality, cost and potential ecological implications. Possible toxic effects are not easily attributed to a certain property of the nanomaterial, or even the nanomaterial itself, since impurities and other components could be held responsible. Therefore, the starting material and its properties must be known, especially for industrially produced ENMs where problems associated with crude production processes have been identified and led to subsequent variations in material properties, e.g., size, shape, etc. In addition, nano-related databases must be consolidated in order to create a definitive reference point. In the meantime, the use of commercial products that contain any nanomaterials should be monitored and regulated.

REFERENCES

Abbasi, E, Aval, SF, Akbarzadeh, A, Milani, M, Nasrabadi, HT, Joo, SW, Hanifehpour, Y, Nejati-Koshki, K & Pashaei-Asl, R. 2014. 'Dendrimers: Synthesis, applications, and properties'. *Nanoscale Research Letters*, 9(1), p 247.

Ahamed, M, AlSalhi, MS & Siddiqui, MKJ. 2010. 'Silver nanoparticle applications and human health'. *Clinica chimica acta*, 411(23), pp 1841–1848.

Ahn, JM, Eom, HJ, Yang, X, Meyer, JN & Choi, J. 2014. 'Comparative toxicity of silver nanoparticles on oxidative stress and DNA damage in the nematode, *Caenorhabditis elegans*'. *Chemosphere*, 108, pp 343–352.

Alarifi, S, Ali, D, Verma, A, Alakhtani, S & Ali, BA. 2013. 'Cytotoxicity and genotoxicity of copper oxide nanoparticles in human skin keratinocytes cells'. *International Journal of Toxicology*, 32(4), pp 296–307.

Alkilany, AM & Murphy, CJ. 2010. 'Toxicity and cellular uptake of gold nanoparticles: What we have learned so far?'. *Journal of Nanoparticle Research*, 12(7), pp 2313–2333.

Allan, J, Reed, S, Bartlett, J & Capra, M. 2009. 'Comparison of methods used to treat nanowaste from research and manufacturing facilities'. Australian Institute of Occupational Hygienists annual conference, Canberra, Australia, pp 5–9. Available at: www.academia.edu/6447281/Comparison_of_methods_used_to_treat_nanowaste_from_research_and_manufacturing_facilities (accessed on 24 February 2015).

Allan, J. 2013. 'Managing nanowaste: Concepts and challenges for nanomanufacturers'. *Nanotechnology Commercialization*, 11, pp 381–402.

Aschberger, K, Micheletti, C, Sokull-Klüttgen, B & Christensen, FM. 2011. 'Analysis of currently available data for characterising the risk of engineered nanomaterials to the environment and human health: Lessons learned from four case studies'. *Environment International*, 37(6), pp 1143–1156.

Auffan, M, Pedeutour, M, Rose, J, Masion, A, Ziarelli, F, Borschneck, D, Chaneac, C, Botta, C, Chaurand, P, Labille, J & Bottero, JY. 2010. 'Structural degradation at the surface of a TiO2-based nanomaterial used in cosmetics'. *Environmental Science & Technology*, 44(7), pp 2689–2694.

Ballestri, M, Baraldi, A, Gatti, AM, Furci, L, Bagni, A, Loria, P, Rapanà, RM, Carulli, N & Albertazzi, A. 2001. 'Liver and kidney foreign bodies granulomatosis in a patient with malocclusion, bruxism, and worn dental prostheses'. *Gastroenterology*, 121(5), pp 1234–1238.

Barroso, MM. 2011. 'Quantum dots in cell biology'. *Journal of Histochemistry & Cytochemistry*, 59(3), pp 237–251.

Bernhardt, ES, Colman, BP, Hochella, MF, Cardinale, BJ, Nisbet, RM, Richardson, CJ & Yin, L. 2010. 'An ecological perspective on nanomaterial impacts in the environment'. *Journal of Environmental Quality*, 39(6), pp 1954–1965.

Binh, CTT, Tong, T, Gaillard, JF, Gray, KA & Kelly, JJ. 2014. 'Acute effects of TiO2 nanomaterials on the viability and taxonomic composition of aquatic bacterial communities assessed via high-throughput screening and next generation sequencing'. *PlOS One*, 9(8), p.e106280.

Boldrin, A, Hansen, SF, Baun, A, Hartmann, NIB & Astrup, TF. 2014. 'Environmental exposure assessment framework for nanoparticles in solid waste'. *Journal of Nanoparticle Research*, 16(6), p 2394.

Brenner, SA & Neu-Baker, NM. 2015. 'Occupational exposure to nanomaterials: Assessing the potential for cutaneous exposure to metal oxide nanoparticles in a semiconductor facility'. *Journal of Chemical Health and Safety*, 22(4), pp 10–19.

Buzea, C, Pacheco, II & Robbie, K. 2007. 'Nanomaterials and nanoparticles:

Sources and toxicity'. *Biointerphases*, 2(4), pp MR17–MR71.

Bystrzejewska-Piotrowska, G, Golimowski, J & Urban, PL. 2009. 'Nanoparticles: Their potential toxicity, waste and environmental management'. *Waste Management*, 29(9), pp 2587–2595.

Campbell, K. 2006. 'SA strategises to exploit nanotech opportunites'. *Engineering News*, 1 May. Available at: www.engineeringnews.co.za/article/sa-strategises-to-exploit-nanotech-opportunites-2006-05-01 (accessed on 27 February 2015).

Chang, X, Fu, Y, Zhang, Y, Tang, M & Wang, B. 2014. 'Effects of Th1 and Th2 cells balance in pulmonary injury induced by nano titanium dioxide'. *Environmental Toxicology and Pharmacology*, 37(1), pp 275–283.

Chen, D & Wang, Y. 2013. 'Impurity doping: A novel strategy for controllable synthesis of functional lanthanide nanomaterials'. *Nanoscale*, 5(11), pp 4621–4637.

Cho, HS, Sung, JH, Song, KS, Kim, JS, Ji, JH, Lee, JH, Ryu, HR, Ahn, K & Yu, IJ. 2013. 'Genotoxicity of silver nanoparticles in lung cells of Sprague Dawley rats after 12 weeks of inhalation exposure'. *Toxics*, 1(1), pp 36–45.

Chung, HE, Yu, J, Baek, M, Lee, JA, Kim, MS, Kim, SH, Maeng, EH, Lee, JK, Jeong, J & Choi, SJ. 2013. 'Toxicokinetics of zinc oxide nanoparticles in rats'. *Journal of Physics: Conference Series*, 429(1), p 012037. IOP Publishing.

Claassens, CH & Motuku, M. 2006. 'Nanoscience and nanotechnology research and development in South Africa'. *Nanotechnology Law and Business*, 3(2). Available at: www.nanolabweb.com/index.cfm/action/main.default.viewArticle/articleID/139/CFID/16564917/CFTOKEN/683b397fb39b2e40-4D1C2022-5056-B577-01685540433D9ACF/index.html (accessed on 24 February 2015).

Colvin, VL. 2003. 'The potential environmental impact of engineered nanomaterials'. *Nature Biotechnology*, 21(10), pp 1166–1170.

Comar Chemicals. 2014. 'Comar Chemicals: Manufacturers of carboxylates'. Available at: www.comarchemicals.com/index.php/en/ (accessed on 22 October 2014).

Crosera, M, Bovenzi, M, Maina, G, Adami, G, Zanette, C, Florio, C & Larese, FF. 2009. 'Nanoparticle dermal absorption and toxicity: A review of the literature'. *International Archives of Occupational and Environmental Health*, 82(9), pp 1043–1055.

CSIR Communication. 2007. 'South Africa launches first nanotechnology innovation centres'. Available at: www.csir.co.za/general_news/2007/first_nano_centres.html (accessed on 22 October 2014).

CSIR Communication. 2009a. 'CSIR research platform to assess risks of nanotechnology'. Available at: www.csir.co.za/enews/2009_may/nre_04.html (accessed on 22 October 2014).

CSIR Communication. 2009b. 'First South African national workshop on nanotechnology risk assessment'. Available at: www.csir.co.za/nre/pollution_and_waste/nano_workshop.html (accessed on 22 October 2014).

De Jager, P, Masoka, X, Sanabria, N, Vetten, M, Andraos, C, Boodhia, K, Koekemoer, L, Matatiele, P, Singh, E, Mabeqa, NT, Mtembu, N & Gulumian, M. 2014. 'Nanotechnology in South Africa: A baseline study'. *A report prepared for Department of Science and Technology*, March 2014, Pretoria, 104 pages.

De Marzi, L, Monaco, A, De Lapuente, J, Ramos, D, Borras, M, Di Gioacchino, M, Santucci, S & Poma, A. 2013. 'Cytotoxicity and genotoxicity of ceria nanoparticles on different cell lines in vitro'. *International Journal of Molecular Sciences*, 14(2), pp 3065–3077.

Di Bucchianico, S, Fabbrizi, MR, Cirillo, S, Uboldi, C. Gilliland, D, Valsami-Jones, E & Migliore, L. 2014. 'Aneuploidogenic effects and DNA oxidation induced in vitro by differently sized gold nanoparticles'. *International Journal of Nanomedicine*, 9, pp 2191–2204.

Donaldson, K, Stone, V, Tran, CL, Kreyling, W & Borm, PJA. 2004. 'Nanotoxicology'. *Occupational and Environmental Medicine*, 61, pp 727–728.

DST Communication. 2008. 'Nanoscience and nanotechnology: 10-year research plan'. *Science and Technology*. Pretoria, South Africa, 28 pages.

DST Communication. 2013. 'Annual report 2012/2013. SA Technology'. 6 pages. Available at: www.gov.za/sites/www.gov.za/files/Department_of_Science_Technology_Annual_Report2012-2013.pdf (accessed on 27 February 2015).

Edwards, J, Harford, A, Wright, P & Priestly, B. 2007. 'Current OHS best practices for the Australian nanotechnology industry: A position paper by the NanoSafe Australia Network'. *Journal of Occupational Health and Safety, Australia and New Zealand*, 23(4), p 315. Available at: www.mams.rmit.edu.au/72nuxiavskpg.pdf (accessed on 27 February 2015).

EFSA Scientific Committee. 2011. 'Guidance on the risk assessment of the application of nanoscience and nanotechnologies in the food and feed chain'. *EFSA Journal*, 9(5), pp 2140. doi:10.2903/j.efsa.2011.2140.

Elder, A, Gelein, R, Silva, V, Feikert, T, Opanashuk, L, Carter, J, Potter, R, Maynard, A, Ito, Y, Finkelstein, J & Oberdörster, G. 2006. 'Translocation of inhaled ultrafine manganese oxide particles to the central nervous system'. *Environmental Health Perspectives*, 114(8), pp 1172–1178.

Elsaesser, A & Howard, CV. 2012. 'Toxicology of nanoparticles'. *Advanced Drug Delivery Reviews*, 64(2), pp 129–137.

El-Sayed, YS, Shimizu, R, Onoda, A, Takeda, K & Umezawa, M. 2015. 'Carbon black nanoparticle exposure during middle and late fetal development induces immune activation in male offspring mice'. *Toxicology*, 327, pp 53–61.

Florence, AT. 2005. 'Nanoparticle uptake by the oral route: Fulfilling its potential?'. *Drug Discovery Today: Technologies*, 2(1), pp 75–81.

George, S, Lin, S, Ji, Z, Thomas, CR, Li, L, Mecklenburg, M, Meng, H, Wang, X, Zhang, H, Xia, T & Hohman, JN. 2012. 'Surface defects on plate-shaped silver nanoparticles contribute to its hazard potential in a fish gill cell line and zebrafish embryos'. *ACS Nano*, 6(5), pp 3745–3759.

Geraets, L, Oomen, AG, Krystek, P, Jacobsen, NR, Wallin, H, Laurentie, M, Verharen, HW, Brandon, EF & De Jong, WH. 2014. 'Tissue distribution and elimination after oral and intravenous administration of different titanium dioxide nanoparticles in rats'. *Part Fibre Toxicology*, 11(1), p 30. doi:10.1186/1743-8977-11-30.

Glista-Baker, EE, Taylor, AJ, Sayers, BC, Thompson, EA & Bonner, JC. 2014. 'Nickel nanoparticles cause exaggerated lung and airway remodeling in mice lacking the T-box transcription factor, TBX21 (T-bet)'. *Particle and Fibre Toxicology*, 11(1), p 7. doi:10.1186/1743-8977-11-7.

Goldman, L & Coussens, C. 2005. 'Implications of nanotechnology for environmental health research'. Institute of Medicine (US) roundtable on environmental health sciences, research, and medicine. Washington (DC): National Academies Press.

Gullapalli, S & Wong, MS. 2011. 'Nanotechnology: A guide to nano-objects'. *Chemical Engineering Progress*, 107(5), pp 28–32.

Gulumian, M. 2009. 'Nanoparticles, nanoscience and nanotechnologies in South Africa'. *UNITAR*, 19 pages. Available at: www2.unitar.org/cwm/publications/event/Nano/Abidjan_25-26_Jan_10/22_South_Africa.pdf (accessed on 27 February 2015).

Gulumian, M, Kuempel, ED & Savolainen, K. 2012. 'Global challenges in the risk assessment of nanomaterials: Relevance to South Africa'. *South African Journal of Science*, 108(9–10), pp 1–9.

Gustafsson, Å, Jonasson, S, Sandström, T, Lorentzen, JC & Bucht, A. 2014. 'Genetic variation influences immune responses in sensitive rats following exposure to TiO 2 nanoparticles'. *Toxicology*, 326, pp 74–85

Han, SG, Lee, JS, Ahn, K, Kim, YS, Kim, JK, Lee, JH, Shin, JH, Jeon, KS, Cho, WS, Song, NW & Gulumian, M. 2015. 'Size-dependent clearance of gold nanoparticles from lungs of Sprague–Dawley rats after short-term inhalation exposure'. *Archives of Toxicology*, 89(7), pp 1083–1094. doi:10.1007/s00204-014-1292-9.

Handy, RD. 2008. 'Systems toxicology: Using the systems biology approach to assess chemical pollutants in the environment'. *Advances in Experimental Biology*, 2, pp 249–281.

Hansen, SF. 2010. 'A global view of regulations affecting nanomaterials'. *Wiley Interdisciplinary Reviews: Nanomedicine and Nanobiotechnology*, 2(5), pp 441–449.

Helmus, MN. 2007. 'The need for rules and regulations'. *Nature*

Nanotechnology, 2, pp 333–334.

Hofkens, J, Maus, M, Gensch, T, Vosch, T, Cotlet, M, Köhn, F, Herrmann, A, Müllen, K & De Schryver, F. 2000. 'Probing photophysical processes in individual multichromophoric dendrimers by single-molecule spectroscopy'. *Journal of the American Chemical Society*, 122(38), pp 9278–9288.

Horie, M, Nishio, K, Kato, H, Fujita, K, Endoh, S, Nakamura, A, Miyauchi, A, Kinugasa, S, Yamamoto, K, Niki, E & Yoshida, Y. 2011. 'Cellular responses induced by cerium oxide nanoparticles: Induction of intracellular calcium level and oxidative stress on culture cells'. *The Journal of Biochemistry*, 150(4), pp 461–471.

Huang, X, Li, L, Liu, T, Hao, N, Liu, H, Chen, D & Tang, F. 2011. 'The shape effect of mesoporous silica nanoparticles on biodistribution, clearance, and biocompatibility *in vivo*'. *ACS Nano*, 5(7), pp 5390–5399.

Hussain, M, Shchepinov, M, Sohail, M, Benter, IF, Hollins, AJ, Southern, EM & Akhtar, S. 2004. 'A novel anionic dendrimer for improved cellular delivery of antisense oligonucleotides'. *Journal of Controlled Release*, 99(1), pp 139–155.

Hussain, S, Thomassen, LC, Ferecatu, I, Borot, MC, Andreau, K, Martens, JA, Fleury, J, Baeza-Squiban, A, Marano, F & Boland, S. 2010. 'Carbon black and titanium dioxide nanoparticles elicit distinct apoptotic pathways in bronchial epithelial cells'. *Particle and Fibre Toxicology*, 7(1), p 10. doi:10.1186/1743-8977-7-10.

Iannuccelli, V, Bertelli, D, Romagnoli, M, Scalia, S, Maretti, E, Sacchetti, F & Leo, E. 2014. '*In vivo* penetration of bare and lipid-coated silica nanoparticles across the human stratum corneum'. *Colloids and Surfaces B: Biointerfaces*, 122, pp 653–661.

ISO. 2010. 'Nanotechnologies – Methodology for the classification and categorization of nanomaterial'. Available at: www.iso.org/iso/catalogue_detail?csnumber=55967 (accessed in July 2016).

Jain, S, Singh, SR & Pillai, S. 2012. 'Toxicity issues related to biomedical applications of carbon nanotubes'. *Journal of Nanomedicine & Nanotechnology*, 3, pp 140–155.

Jani, PU, Florence, AT & McCarthy, DE. 1992. 'Further histological evidence of the gastrointestinal absorption of polystyrene nanospheres in the rat'. *International Journal of Pharmaceutics*, 84(3), pp 245–252.

Jiang, J, Oberdörster, G & Biswas, P. 2009. 'Characterization of size, surface charge, and agglomeration state of nanoparticle dispersions for toxicological studies'. *Journal of Nanoparticle Research*, 11(1), pp 77–89.

Ju-Nam, Y & Lead, JR. 2008. 'Manufactured nanoparticles: An overview of their chemistry, interactions and potential environmental implications'. *Science of the Total Environment*, 400(1), pp 396–414.

Kang, GS, Gillespie, PA, Gunnison, A, Moreira, AL, Tchou-Wong, KM &

Chen, LC. 2011. 'Long-term inhalation exposure to nickel nanoparticles exacerbated atherosclerosis in a susceptible mouse model'. *Environmental Health Perspectives*, 119(2), pp 176–178.

Khan, FR, Kennaway, GM, Croteau, MN, Dybowska, A, Smith, BD, Nogueira, AJ, Rainbow, PS, Luoma, SN & Valsami-Jones, E. 2014. '*In vivo* retention of ingested AuNPs by Daphnia magna: No evidence for trans-epithelial alimentary uptake'. *Chemosphere*, 100, pp 97–104.

Kim, B, Kim, H & Yu, IJ. 2014. 'Assessment of nanoparticle exposure in nanosilica handling process: Including characteristics of nanoparticles leaking from a vacuum cleaner'. *Industrial Health*, 52(2), pp 152–162.

Kim, JS, Sung, JH, Ji, JH, Song, KS, Lee, JH, Kang, CS & Yu, IJ. 2011. '*In vivo* genotoxicity of silver nanoparticles after 90-day silver nanoparticle inhalation exposure'. *Safety and Health at Work*, 2(1), pp 34–38.

Köhler, AR, Som, C, Helland, A, Gottschalk, F. 2008. 'Studying the potential release of carbon nanotubes throughout the application life cycle'. *Journal of Cleaner Production*, 16(8–9), pp 927–937.

Kreider, ML, Burns, AM, DeRose, GH & Panko, JM. 2013. 'Protecting workers from risks associated with nanomaterials: Part II, best practices in risk management'. *Occupational Health & Safety (Waco, Tex.)*, 82(9), pp 20–24.

Kumar, A & Dhawan, A. 2013. 'Genotoxic and carcinogenic potential of engineered nanoparticles: An update'. *Archives of Toxicology*, 87(11), pp 1883–1900.

Lee, JH, Kim, YS, Song, KS, Ryu, HR, Sung, JH, Park, JD, Park, HM, Song, NW, Shin, BS, Marshak, D & Ahn, K. 2013. 'Biopersistence of silver nanoparticles in tissues from Sprague–Dawley rats'. *Particle and Fibre Toxicology*, 10(1), p 36. doi:10.1186/1743-8977-10-36.

Leite-Silva, VR, Le Lamer, M, Sanchez, WY, Liu, DC, Sanchez, WH, Morrow, I, Martin, D, Silva, HD, Prow, TW, Grice, JE & Roberts, MS. 2013. 'The effect of formulation on the penetration of coated and uncoated zinc oxide nanoparticles into the viable epidermis of human skin in vivo'. *European Journal of Pharmaceutics and Biopharmaceutics*, 84(2), pp 297–308.

Lewinski, N, Colvin, Drezek, R. 2008. 'Cytotoxicity of nanoparticles'. *Small*, 4(1), pp 26–49.

Ma-Hock, L, Farias, PMA, Hofmann, T, Andrade, ACDS, Silva, JN, Arnaud, TMS, Wohlleben, W, Strauss, V, Treumann, S, Chaves, CR & Gröters, S. 2014. 'Short term inhalation toxicity of a liquid aerosol of glutaraldehyde-coated CdS/Cd (OH) 2 core shell quantum dots in rats'. *Toxicology Letters*, 225(1), pp 20–26.

Manke, A, Wang, L, Rojanasakul, Y. 2013. 'Mechanisms of nanoparticle-induced oxidative stress and toxicity. *BioMed Research International*. doi.org/10.1155/2013/942916.

Martirosyan, A & Schneider, YJ. 2014. 'Engineered nanomaterials in food:

Implications for food safety and consumer health'. *International Journal of Environmental Research and Public Health*, 11(6), pp 5720–5750.

Maynard, AD. 2006. 'Nanotechnology: Assessing the risks'. *Nano Today*, 1(2), pp 22–33.

Maynard, AD. 2008. 'Living with nanoparticles'. *Nano Today*, 3, pp 64–64.

Maynard, AD, Aitken, RJ, Butz, T, Colvin, V, Donaldson, K, Oberdörster, G, Philbert, MA, Ryan, J, Seaton, A, Stone, V & Tinkle, SS. 2006. 'Safe handling of nanotechnology'. *Nature*, 444(7117), pp 267–269.

McLean, M. 2000. 'Informal collection: A matter of survival amongst the urban vulnerable'. *Africanus*, 30(2), pp 8–26.

McNeil, SE. 2005. 'Nanotechnology for the biologist'. *Journal of Leukocyte Biology*, 78(3), pp 585–594.

Medina, M. 2010. 'Solid wastes, poverty and the environment in developing country cities: Challenges and opportunities'. Working paper: World Institute for Development Economics Research. Available at: www.wider.unu.edu/publications/working-papers/2010/en_GB/wp2010-23/ (accessed on 27 February 2015).

MINTEK. 2014. 'Nanotechnology: Advanced materials'. Available at: www.mintek.co.za/technical-divisions/advanced-materials-amd/nanotechnology/ (accessed on 22 October 2014).

Mittal, S & Pandey, AK. 2014. 'Cerium oxide nanoparticles induced toxicity in human lung cells: Role of ROS mediated DNA damage and apoptosi'. *BioMed Research International*. doi.org/10.1155/2014/891934.

Monteiro-Riviere, NA & Inman, AO. 2006. 'Challenges for assessing carbon nanomaterial toxicity to the skin'. *Carbon*, 44(6), pp 1070–1078.

Musee, N. 2011. 'Nanotechnology risk assessment from a waste management perspective: Are the current tools adequate?'. *Human & Experimental Toxicology*, 30(8), pp 820–835.

Nano Bubble Technology. 2014. 'Nano Bubble Technology (NBT): A revolution in green fuel technologies'. Available at: www.nanobubbletech.com/index.php (accessed on 22 October 2014).

NanoBugle. 2012. 'The "Rule-of-Three". NanoTech (South Africa). Nanotechnology companies: South Africa'. Available at: www.nanobugle.org/tag/nanotechnology-companies-south-africa/ (accessed on 22 October 2014).

Nanotech Energy. 2014. 'Renewable Energy Products'. Available at: www.nanotechafrica.co.za/index.html (accessed on 22 October 2014).

NanoTech Water Solutions. 2014. 'Water purification and water treatment – complete turnkey solutions – Guaranteed!' Available at: www.nano-tech.co.za/index.php (accessed on 22 October 2014).

Napierska, D, Thomassen, LC, Lison, D, Martens, JA & Hoet, PH. 2010. 'The nanosilica hazard: Another variable entity'. *Particle and Fibre Toxicology*, 7(1), p 39. doi:10.1186/1743-8977-7-39.

NPEP. 2014. 'Nanotechnology: A world of possibilities'. Available at: www.

npep.co.za/wp-content/uploads/2017/04/npep_fact_sheet_careers_01.pdf. (accessed on 22 October 2014).

NRF. 2014. 'Infrastructure funding instruments: National Equipment Programme (NEP) and National Nanotechnology Equipment Programme (NNEP)'. *NEP & NNEP Strategic Framework 2014/15*, 6 pages. Available at: www.nrf.ac.za/sites/default/files/documents/NEP_NNEP%20_2014_15%20_Framework_May%202014_FINAL%20%282%292.pdf (accessed on 22 October 2014).

Nuñez-Anita, RE, Acosta-Torres, LS, Vilar-Pineda, J, Martínez-Espinosa, JC, De la Fuente-Hernández, J & Castaño, VM. 2014. 'Toxicology of antimicrobial nanoparticles for prosthetic devices'. *International Journal of Nanomedicine*, 9, pp 3999–4006.

Nyokong, T, & Limson, J. 2013. 'An education in progress'. *Nature Nanotechnology*, 8(11), pp 789–791.

Oberdörster, G. 2010. 'Safety assessment for nanotechnology and nanomedicine: Concepts of nanotoxicology'. *Journal of Internal Medicine*, 267(1), pp 89–105.

Oberdörster, G, Oberdörster, E & Oberdörster, J. 2005. 'Nanotoxicology: An emerging discipline evolving from studies of ultrafine particles'. *Environmental Health Perspectives*, 113(7), p 823.

Oelofse, Z, & Musee, N. 2008. 'Emerging issues paper: Hazardous & new waste types. Hazardous waste management & emerging waste streams: A consideration of key emerging issues that may impact the state of the environment'. Available at: www.sludgenews.org/resourcesf (accessed on 12 November 2014).

Osmond, MJ & McCall, MJ. 2010. 'Zinc oxide nanoparticles in modern sunscreens: An analysis of potential exposure and hazard'. *Nanotoxicology*, 4(1), pp 15–41.

Paek, HJ, Lee, YJ, Chung, HE, Yoo, NH, Lee, JA, Kim, MK, Lee, JK, Jeong, J & Choi, SJ. 2013. 'Modulation of the pharmacokinetics of zinc oxide nanoparticles and their fates in vivo'. *Nanoscale*, 5(23), pp 11416–11427.

Park, B, Donaldson, K, Duffin, R, Tran, L, Kelly, F, Mudway, I, Morin, JP, Guest, R, Jenkinson, P, Samaras, Z & Giannouli, M. 2008. 'Hazard and risk assessment of a nanoparticulate cerium oxide-based diesel fuel additive: A case study'. *Inhalation Toxicology*, 20(6), pp 547–566.

Park, K. 2013. 'Toxicokinetic differences and toxicities of silver nanoparticles and silver ions in rats after single oral administration'. *Journal of Toxicology and Environmental Health, Part A*, 76(22), pp 1246–1260.

Peira, E, Turci, F, Corazzari, I, Chirio, D, Battaglia, L, Fubini, B & Gallarate, M. 2014. 'The influence of surface charge and photo-reactivity on skin-permeation enhancer property of nano-TiO2 in ex vivo pig skin model under indoor light'. *International Journal of Pharmaceutics*, 467(2014), pp 90–99.

Quintana, A, Raczka, E, Piehler, L, Lee, I, Myc, A, Majoros, I, Patri, AK, Thomas, T, Mulé, J & Baker, JR. 2002. 'Design and function of a dendrimer-based therapeutic nanodevice targeted to tumor cells through the folate receptor'. *Pharmaceutical Research*, 19(9), pp 1310–1316.

Rahi, A, Sattarahmady, N & Heli, H. 2014. 'Toxicity of Nanomaterials-Physicochemical Effects'. *Austin Journal of Nanomedicine & Nanotechnology*, 2(6), pp 1034–1042.

Rim, KT, Song, SW & Kim, HY. 2013. 'Oxidative DNA damage from nanoparticle exposure and its application to workers' health: A literature review'. *Safety and Health at Work*, 4(4), pp 177–186.

Rouse, JG, Yang, J, Ryman-Rasmussen, JP, Barron, AR & Monteiro-Riviere, NA. 2007. 'Effects of mechanical flexion on the penetration of fullerene amino acid-derivatized peptide nanoparticles through skin'. *Nano Letters*, 7(1), pp 155–160.

Rubber Nano Products. 2014. 'Environmentally superior rubber for industrial gain'. Available at: www.rubbernano.co.za/ (accessed on 22 October 2014).

Ryman-Rasmussen, JP, Riviere, JE & Monteiro-Riviere, NA. 2006. 'Penetration of intact skin by quantum dots with diverse physicochemical properties'. *Toxicological Sciences*, 91(1), pp 159–165.

Ryman-Rasmussen, JP, Riviere, JE & Monteiro-Riviere, NA. 2007. 'Surface coatings determine cytotoxicity and irritation potential of quantum dot nanoparticles in epidermal keratinocytes'. *Journal of Investigative Dermatology*, 127(1), pp 143–153.

SAASTA. 2001. 'Nanotechnology Public Engagement Programme (NPEP)'. Available at: www.saasta.ac.za/projects/nanotechnology-public-engagement-programme-npep (accessed on 24 February 2015).

SAASTA. 2014. 'Nanotechnology Public Engagement Programme (NPEP)'. Available at: www.saasta.ac.za/index.php?option=com_content&view=article&id=75&Itemid=65 (accessed on 22 October 2014).

Saptarshi, SR, Duschl, A & Lopata, AL. 2013. 'Interaction of nanoparticles with proteins: Relation to bio-reactivity of the nanoparticle'. *Journal of Nanobiotechnology*, 11(1), p 26. doi:10.1186/1477-3155-11-26.

Saquib, Q, Al-Khedhairy, AA, Siddiqui, MA, Abou-Tarboush, FM, Azam, A & Musarrat, J. 2012. 'Titanium dioxide nanoparticles induced cytotoxicity, oxidative stress and DNA damage in human amnion epithelial (WISH) cells'. *Toxicology in Vitro,* 26, pp 351–361.

Sargent, LM, Porter, DW, Staska, LM, Hubbs, AF, Lowry, DT, Battelli, L, Siegrist, KJ, Kashon, ML, Mercer, RR, Bauer, AK, Chen, BT, Salisbury, JL, Frazer, D, McKinney, W, Andrew, M, Tsuruoka, S, Endo, M, Fluharty, KL, Castranova, V & Reynolds, SH. 2014. 'Promotion of lung adenocarcinoma following inhalation exposure to multi-walled carbon nanotubes'. *Particle and Fibre Toxicology,* 11, p 3. doi:10.1186/1743-8977-11-3.

SCENIHR. 2009. 'Risk assessment of products of Nanotechnologies Brussels

European Commission: Directorate General for Health and Consumers'. Available at: www.ec.europa.eu/health/ph_risk/committees/04_scenihr/docs/scenihr_o_023.pdf (accessed on 27 February 2015).

Schenck, R & Blaauw, PF. 2011. 'The work and lives of street waste pickers in Pretoria: A case study of recycling in South Africa's urban informal economy'. *Urban Forum*, 22(4), pp 411–430.

Shinde, SK, Grampurohit, ND, Gaikwad, DD, Jadhav, SL, Gadhave, MV & Shelke, PK. 2012. 'Toxicity induced by nanoparticles'. *Asian Pacific Journal of Tropical Disease*, 2(4), pp 331–334.

Shvedova, A, Tkach, AV, Kisin, ER, Khaliullin, T, Stanley, S, Gutkin, DW, Star, A, Chen, Y, Shurin GV, Kagan, VE & Shurin, MR. 2013. 'Carbon nanotubes enhance metastatic growth of lung carcinoma via up-regulation of myeloid-derived suppressor cells'. *Small* 9, pp 1691–1695.

Shvedova, AA, Kisin, ER, Murray, AR, Mouithys-Mickalad, A, Stadler, K, Mason, RP & Kadiiska, M. 2014a. 'ESR evidence for in vivo formation of free radicals in tissue of mice exposed to single-walled carbon nanotubes'. *Free Radical Biology and Medicine*, 73, pp 154–65.

Shvedova, AA, Yanamala, N, Kisin, ER, Tkach, AV, Murray, AR, Hubbs, A, Chirila, MM, Keohavong, P, Sycheva, LP, Kagan, VE & Castranova, V. 2014b. 'Long-term effects of carbon containing engineered nanomaterials and asbestos in the lung: one year post-exposure comparisons'. *American Journal of Physiology-Lung Cellular and Molecular Physiology,* 306, L170-L182.

Simonet BM & Valcárcel M. 2009. 'Monitoring nanoparticles in the environment'. *Analytical and Bioanalytical Chemistry,* 393, pp 17–21.

Singh, N, Manshian, B, Jenkins, GJ, Griffiths, SM, Williams, PM, Maffeis, TG, Wright, CJ & Doak SH. 2009. Nano genotoxicology: The DNA damaging potential of engineered nanomaterials. *Biomaterials,* 30, pp 3891–3914.

Stern, ST & McNeil, SE. 2008. 'Nanotechnology safety concerns revisited'. *Toxicological Sciences,* 101, pp 4–21.

Stewart, GM & Fox, MA. 1996. 'Chromophore-labeled dendrons as light harvesting antennae'. *Journal of the American Chemical Society,* 118, pp 4354–4360.

Takenaka, S, Karg, E, Roth, C, Schulz, H, Ziesenis, A, Heinzmann, U, Schramel, P & Heyder, J. 2001. 'Pulmonary and systemic distribution of inhaled ultrafine silver particles in rats'. *Environmental Health Perspectives,* 109, pp 547–551.

Tan, M-H, Commens, CA, Burnett, L & Snitch, PJ. 1996. 'A pilot study on the percutaneous absorption of microfine titanium dioxide from sunscreens'. *Australasian Journal of Dermatology,* 37, pp 185–187.

Themelis, N, Castaldi, M, Bhatti, J & Arsova, L. 2011. 'Energy and economic value of non-recycled plastics (NRP) and municipal solid wastes (MSW) that are currently landfilled in the fifty states'. Available at: www.seas.

columbia.edu/earth/wtert/sofos/ACC_Final_Report_August23_2011.pdf (accessed on 27 February 2015).

Tsuji, JS, Maynard, AD, Howard, PC, James, JT, Lam, CW, Warheit, B & Santamaria, AB. 2006. 'Research strategies for safety evaluation of nanomaterials, Part IV: Risk assessment of nanoparticles'. *Toxicological Sciences,* 89, pp 42–50.

UNISA. 2014. 'Dynamite comes in nano packages'. Available at: www.unisa.ac.za/news/index.php/2013/05/dynamite-comes-in-nano-packages/ (accessed on 22 October 2014).

Unrine, JM, Colman, BP, Bone, AJ, Gondikas, AP & Matson, CW. 2012. 'Biotic and abiotic interactions in aquatic microcosms determine fate and toxicity of Ag nanoparticles. Part 1: Aggregation and dissolution'. *Environmental Science & Technology,* 46, pp 6915–6924.

Van der Zande, M, Vandebriel, RJ, Van Doren, E, Kramer, E, Herrera Rivera, Z, Serrano-Rojero, CS, Gremmer, ER, Mast, J, Peters, RJ, Hollman, PC, Hendriksen, PJ, Marvin, HJ, Peijnenburg, AA & Bouwmeester, H. 2012. 'Distribution, elimination, and toxicity of silver nanoparticles and silver ions in rats after 28-day oral exposure'. *ACS Nano,* 6, pp 7427–7442.

Vandebriel, RJ, Tonk, EC, De la Fonteyne-Blankestijn, LJ, Gremmer, ER, Verharen, HW, Van der Ven, LT, Van Loveren, H & de Jong, WH. 2014. 'Immunotoxicity of silver nanoparticles in an intravenous 28-day repeated-dose toxicity study in rats'. *Particle and Fibre Toxicology,* doi:10.1186/1743-8977-11-21.

Vanderveen, C. 2014. 'Beyond recycling: Recovering the energy in non-recycled plastics'. Available at: www.govtech.com/fs/news/Beyond-Recycling-Recovering-the-Energy-in-Non-Recycled-Plastics-.html (accessed on 26 February 2015).

Vlachogianni, T, Fiotakis, K, Loridas, S, Perdicaris, S & Valavanidis, A. 2013. 'Potential toxicity and safety evaluation of nanomaterials for the respiratory system and lung cancer'. *Lung Cancer: Targets and Therapy,* 4, pp 71–82.

Wild, S. 2013. 'Researchers try get a glimpse of nanotechnology's potential risks. National: science & environment'. Available at: www.bdlive.co.za/national/science/2013/01/24/researchers-try-get-a-glimpse-of-nanotechnologys-potential-risks (accessed on 24 February 2015).

Wilson, DC, Velis, C & Cheeseman, C. 2006. 'Role of informal sector recycling in waste management in developing countries'. *Habitat International,* 30(4), pp 797–808.

Wu, Q, Nouara, A, Li, Y, Zhang, M, Wang, W, Tang, M, Ye, B, Ding, J & Wang, D. 2013. 'Comparison of toxicities from three metal oxide nanoparticles at environmental relevant concentrations in nematode *Caenorhabditis elegans*'. *Chemosphere,* 90, pp 1123–31.

Xia, X, Monteiro-Riviere, NA & Riviere, JE. 2010. 'Skin penetration and kinetics of pristine fullerenes (C60) topically exposed in industrial organic solvents'. *Toxicology and Applied Pharmacology*, 242, pp 29–37.

Yang, H & Lopina, ST. 2003. 'Penicillin V-conjugated PEG-PAMAM star polymers'. *Journal of Biomaterials Science, Polymer*, 14, pp 1043–1056.

Yokel, RA & MacPhail, RC. 2011. 'Engineered nanomaterials: Exposures, hazards, and risk prevention'. *Journal of Occupational Medicine and Toxicology*, 6, p 7. doi:10.1186/1745-6673-6-7.

Yu, KN, Yoon, TJ, Minai-Tehrani, A, Kim, JE, Park, SJ, Jeong, MS, Ha, SW, Lee, JK, Kim, JS & Cho, MH. 2013. 'Zinc oxide nanoparticle induced autophagic cell death and mitochondrial damage via reactive oxygen species generation'. *Toxicology in Vitro*, 27, pp 1187–1195.

THREE

Envisioning and engaging the societal implications of nanotechnology

Is it too early for Africa to do nanoethics?

HAILEMICHAEL T DEMISSIE

INTRODUCTION

THE SPEED AND ABRUPTNESS of developments in science and technology are of such import that humanity could be caught completely unprepared for the changes brought about by it. With nanotechnology as its catalyst, 'the era of exponential change is fast upon us'. Hardly a day passes without news of some techno-scientific discovery or breakthrough. It is now a safe bet that the promises and dangers of nanotechnology are not only possible but also more imminent than one might think.

An intense debate is underway on whether we should wait for the nanotechnology-future before identifying and addressing the ethical issues associated with it or whether we should start exploring ethical issues and attempt to resolve them in anticipation of the future that is being shaped by current advances. The trajectory of nanotechnology and the technological determinism evident in current trends indicate that deferring the ethical issues would entail a colossal opportunity cost. This is because '[t]he most powerful point at which ethics can have an influence is while the sociotechnical systems of nanotechnology are being envisioned and constructed' (Johnson, 2007: 27). It is also advised that '[a]t the very least we need to do more to be more proactive and less reactive in doing ethics' (Moor, 2005: 119).

Given Africa's troubled encounter with previous technologies that were deployed to facilitate slavery, colonialism and apartheid, the need to engage the ethics of nanotechnology cannot be overemphasised. However, there is scant research on the issue and a gaping void marks the research on Africa's engagement, not only with the ethics of nanotechnology, but also with nanotechnology in general. Ugandan President Yoweri Museveni was right to call upon Africa to move on to the debate on nanotechnology based on his observation that Africa remains bogged in the GM (genetic modification) debate while the rest of the world graduated to the debate on nanotechnology and other emerging technologies (Wamboga-Mugirya, 2008). The need to move on to the debate on the ethics of emerging technologies is further accentuated by the rate of technological advances:

> *There is a deeper, more complex problem associated with the accelerating rate of development. We are already approaching a stage at which ethical issues are emerging, one upon another, at a rate that outstrips our capacity to think through and appropriately respond* (Khushf, 2006: 258).

This chapter provides a comprehensive overview of the debate on the global ethics of nanotechnology or nanoethics as it came to be known. It will explore how Africa is engaging with this debate and will focus on the phenomenon of the 'nano-divide' also known as 'nano-apartheid' that is evolving in the trails of the digital divide. Africa has been the face of all things negative when it comes to the digital divide. Will it be in the same predicament vis-a-vis nanotechnology remaining on the 'bleeding' rather than the leading edge of technology? What policy considerations, regulatory measures and political decisions will be required to ensure that Africa will not miss out on the nanotechnology revolution as it did with previous technologies? Such issues need to be addressed as ethical issues first. It is ethics that lay the groundwork for political actions, policy and regulatory measures.

THE GLOBAL NANOETHICS DEBATE: DEFINING THE ESSENCE AND SCOPE OF NANOETHICS

With a phenomenal outpouring of literature and avid interest from several directions, nanoethics as a distinct discipline is surely gaining momentum and recognition. Like the technology it deals with, nanoethics is not only born but 'a baby that's all growed [sic] up' to borrow the words of Nobel laureate nano-pioneer, Richard Smalley (Kahn, 2006). Nanoethics is indeed very much alive and kicking with no sign of 'birth pangs' as some have tried to portray it (Parens et al, 2008: 1449). Its issues include those dealt with in the context of preceding technologies. What makes an ethical approach to it distinct from preceding ethical disciplines is its potential to correct and consolidate previous approaches and, above all, introduce new approaches and themes. The range of approaches adopted in nanoethics includes but is not limited to ethical, philosophical, theological, historical, anthropological, sociological, political and legal approaches.

Nanoethics is for the moment the catch-all neologism capturing 'the full spectrum of non-technical issues related to nanotechnology' (Cameron & Mitchell, 2007: 362). The description by Lin and Allhoff gives an even wider scope of nanoethics: 'nanoethics means something like the ethical, social, environmental, medical, political, economic, legal issues and so on, arising from nanotechnology' (Lin & Allhoff, 2007: 10; see also Cameron, 2007a: 284). Even the technical issues of the basic science itself, for example, the scientific analysis of environmental impact studies seem to be entangled in the dauntingly broad nanoethics net.

A flurry of competing acronyms has emerged to accommodate the various aspects of the ethics on nanotechnology: these include NELSI (ELSI on nanotechnology) – the Ethical, Legal Social Implications

(ELSI); ELSA (Ethical, Legal Social Aspects), SEIN (social ethical implications/interactions of nanotechnology), NE^3LS (nano-ethical, environmental, economic and legal and social issues), and OELSI (other ELSI). Nanoethics refers to an area of inquiry that explores legal issues as explicitly as the ethical aspects of nanotechnology. It also encroaches upon the realm of the relatively long-established discipline of bioethics. Bioethics is competing with nanoethics in the fast-burgeoning field of nanomedicine (Cameron, 2007a: 285).

Surely the scope of nanoethics is unmanageably broad and arguably 'not yet in focus' (Sandler, 2007: 447). A good deal of the criticism of nanoethics takes advantage of this poorly defined scope. The unceasing expansion of the neologism has been met with criticism by those who warn nanoethics protagonists about a backlash against a supposedly premature ethical engagement with the nascent technology (Caplan, 2008; Keiper, 2007). Whether nanotechnology is ripe for ethical analysis and engagement is not a settled issue and there is considerable ambivalence about when to engage with nanoethics. The 2006 US National Research Council review of the National Nanotechnology Initiative recommends that it is 'not too soon' to engage with the ethical issues now while stressing in its assessment that 'near-term and tangible ethical concerns related to the use of nanotechnology have yet to be determined' (Sandler, 2007: 447). Without naming its task as nanoethics, UNESCO is proceeding with the job, emphasising that there should be no waiting for triggers from either the public or from developments in the technology (Ten Have, 2007: 13).

Nanoethics detractors do not discard the idea in its entirety but appeal for patience until the need for nanoethics becomes sufficiently evident (Keiper, 2007; Litton, 2007). The detractors denouncing the formation of a distinct discipline of nanoethics seem to base their opposition on the idea that this discipline is premature. They would like to see nanotechnology driving ethics and thus would like to postpone any ethical debates for a later point in time. The conclusion reached by the sceptics who maintain the importance of keeping a premature nanoethics off the debate is simple: 'Its time may come someday, but it is too soon to say just when and how' (Keiper, 2007; see also Caplan, 2008: 42). This is the background that must be borne in mind when the question arises as to whether it is too early for Africa to engage in nanoethics.

The temptation to stay in the comfort of familiar paradigms in the face of an emergent field is a natural stage in the evolution of a new discipline. This is explained by the so-called 'emergence theory'. The new enwombed in the features from the old combines them with its own new features to evolve into a whole greater than the sum of its parts (Chesters & Welsh, 2005: 192; Allhoff & Lin, 2006). It is hard for sceptics to dispute the fact that nanotechnology has already evolved into such a gestalt (Kaiser, 2006: 667). By the same token, nanoethics detractors would find it increasingly difficult to deny a corresponding gestalt manifesting in a new field of ethical inquiry. As was the case with bioethics, the emergence of nanotechnology, and correspondingly of nanoethics, as new disciplines is becoming more apparent by the day.

With the controversy largely reduced to the question of timing, the debate on the desirability of nanoethics is not as polarised as it appears. Those critical of nanoethics seem to be mainly concerned with efficiency. Are we not wasting resources by delving into the ethics of unknown and distant developments? Should we pay heed to the argument that it is premature to debate the ethics of nanotechnology and we should rather wait until any dangers are realised? Certainly much of the discipline of the ethics of nanotechnology will concern future developments. However, these developments are disarmingly closer than is often thought.

There are compelling signs that developments thus far point to future certainties rather than contingent possibilities of unknown likelihood. There is a particularly relevant historical precedent from biotechnology where many supposed impossibilities were turned into achievements 'while the early naysayers were still alive' (Silver, 1997: 288). The ideas behind the Drexlerian concept of molecular manufacturing, human enhancement, the space elevator or even the dreaded 'grey goo' are all plausible and cannot be ruled out as irrational. The view that the promises and dangers of nanotechnology are more imminent than they appear is constantly being validated by the developments in nanotechnology.

Neither the wait-and-see attitude of those who dismiss nanoethics as a discipline nor the assurances that pre-nanoethics will suffice to handle nanotechnology can dispense with the need for an ethical discourse on the opportunities and challenges being posed by nanotechnology. Dedicated engagement is required to tackle the ethical issues relating to nanotechnology. What Hans Jonas has said in the general context of contemporary technoscience is particularly true of the ethics of nanotechnology: '[our contemporary technoscience] calls upon the utter resources of ethical thought' (Jonas, 1984: 17). Perhaps his quotation is even more applicable to nanotechnology: Reckoning the demand for a whole new ethics required to deal with the consequences of the fast developing biotechnologies, Dupuy observes that we are fast running out of ethical resources in an inverse proportion to our technological advance (Dupuy, 2008: 254). Stretching our existing ethical norms to govern advances in nanotechnology is unlikely to be a successful undertaking. Unforeseen ethical issues are to crop up with the unprecedented realities that nanotechnology in convergence with other technologies will be creating.

Is this concern relevant to Africa? Should Africa be concerned about nanotechnology forging full steam ahead irrespective of the implications it may have on Africa? Are there historical precedents pointing to the fact that it would be wise to hold the ethics debate sooner rather than later? These and other questions will certainly require the attention of ethicists especially those taking an African perspective on the ethics of technology.

ARE THERE NEW ETHICAL ISSUES WITH NANOTECHNOLOGY?

The argument that nanotechnology does not raise issues that have not been presented by earlier technologies is hardly tenable even for the present stage of the technology (McGinn, 2010: 120). Nanotechnology arrived at a unique historical juncture in the wake of a great deal of institutional change that resulted from preceding technological revolutions, especially the biotech revolution. The shift from academic to post-academic science epitomised by the Bayh-Dole Act in the US, and globalised through the rules of the World Trade Organization (WTO), is a typical example of the change in the environment in which a particular technology is pursued (Moore, 2006: 156). Nanotechnology is unique even among post-academic science as it symbolises 'the first full embodiment of post-academic science' (Vogt et al, 2007: 330; Moriarty, 2008: 60).

The questions we ask about nanotechnology are bound to be new, as although they are raised in a setting engendered by preceding technologies, these technologies never ceased evolving. Even in cases where the questions remain unchanged, it is likely that the answers are different and, in turn, these new answers give rise to new questions. This is the dynamics of the dialectical interaction between ethics and technology.[5] Lin and Allhof's use of the Sorites Paradox also known as 'The Paradox of the Heap' is helpful here (Allhoff & Lin, 2008: 49). Would a heap of sand remain a heap if we remove scoops of sand from the heap? How much sand should there be for a heap of sand to remain a heap? The riddle can be extended also by asking when a heap turns into a hill or a mountain. What we take away from and add on existing ethics will define the emergence of a new ethics.

If new elements and perspectives are introduced to an existing ethics portfolio – which is certainly happening in relation to nanotechnology – the ethics will not remain the same. The hybridity resulting from the convergence of technologies with nanotechnology at their centre entails a combination of features and properties drawn from various technologies. Existing precedents may not be fit to govern the hybrid result; the effectiveness of any overstretching of existing ethical discourses in this respect is highly questionable as there will be 'some striking differences between the debate then and now' (Khushf, 2007a: 308).

5 Bennett-Woods's analysis of utilitarianism in the context of nanotechnology is instructive in this respect. She draws on the common argument that the technology is too new to reliably predict the consequences limiting the application of established utilitarian principles (Bennett-Woods, 2008: 75).

BEYOND ETHICS: THE PEDAGOGY AND THE METHOD OF NANOETHICS

The 'nothing-new-under-the-sun' wisdom detected in the views of nanoethics detractors is reminiscent of the grave errors of the pre-post-modern thinkers who came up with their grand narratives about final paradigms. The great American philosopher Charles Peirce indicts any philosophy that seeks or tends to block inquiry as guilty of 'the one unpardonable offence in reasoning'. This 'venomous error' takes multiple forms. Pierce lists the familiar ones (in Schwandt, 1990: 258):

> *absolute assertion, the claim that certain things can never be known; the claim that certain elements of science are basic, ultimate, independent of all else and utterly inexplicable; and the claim that a particular law or truth has found its last and perfect formulation.*

The claims of nanoethics detractors take one or more of these forms too. The general import of their message is to reassert the fundamental nature, the exhaustive scope and the possibly immutable status of the ethics of the last four decades. Their argument essentially tells us that it is the end of the road for those who wish to ask questions or that we are left with only few questions to ask. Nanoethics sceptics cannot hope to engage with ethics while claiming that there are no or few questions to ask as ethics entails asking questions and then asking nfurther questions. And as Bennett-Woods (2008: 72) argues in the nanotechnology context, failing to ask the questions or asking too few of them is 'a failure of critical thought'.

The heuristics of nanoethics detractors constrain and dampen a rather vibrant debate that uniquely brought together varied forms of scholarship, and in so doing revived hopes of a happy ending to what may be termed the tale of two towers: one of 'ivory' – technoscience research insulated from social concerns – and one of 'Babel' in which specialists talk past each other[6] (Vogt et al, 2007: 329; Khushf, 2007b: 186;

6 Hunt speaks of the need to deal with the modern-day Tower of Babel which Vogt and others term 'the Tower of Babel syndrome' (Vogt et al, 2007: 329; Khushf, 2007b: 186; Hunt, 2008: 275).

Hunt, 2008: 275). Like bioethics before it, nanoethics is to be the market place where thinkers and theorists from all walks of life will meet. What O'Neill (2002: 1) says about bioethics is even more valid for nanoethics:

> *Bioethics is not a discipline ... It has become a meeting ground for a number of disciplines, discourses, and organisations concerned with ethical, legal, and social questions raised by advances in medicine, science and biotechnology.*

Furthermore, the legacy of ethics that nanoethics detractors pride themselves on is not capable of carrying out the mission they seek to reserve for it. The post-modern blow to the grand narratives of modernity has left us with a shabby ethics. This is inadvertently acknowledged by an arch-opponent of the idea of nanoethics, Adam Keiper (2007). He sees nanoethics as the lazy way out from the post-modern epistemic quagmire that impaired our ability to identify ethical principles.

Taking stock of the legacy ethics of today reveals a disturbing inventory. Alasdair MacIntyre's damning diagnosis of this state of affairs is paraphrased in these startling words:

> *our moral culture is in a grave state of disorder lacking any comprehensive and coherent understanding of morality and human nature, we subsist on scattered shards and remnants of past moral frameworks* (MacIntyre, 1985: 253; see also Arras, 1997: 179).

In the narrower context of technoscience, Somerville appeals for ethics that will trade the shards of past moral stories for a 'shared story' (Somerville, 2004: 5).

Cognisant of the tendency to block the incoming nanoethics discourse, Hunt and Mehta (2006: 7) stress the frame of mind that those engaging with the debate on nanotechnology should be adopting. They look forward to anticipating whether a greater degree of maturity will be exhibited by stakeholders in forthcoming nanotechnology disagreements. This maturity would include a willingness to respond to new ideas and tentative findings with open-mindedness, inquisitiveness, constructive suggestions and refusal to uphold sectarian interests.

Those who deny the emergence of new issues or assert that the received ethics we have will suffice for the analysis, evaluation and governance of nanotechnology cannot claim to have acquired the open-mindedness that Hunt and Mehta (2006) rightly emphasised as one of 'the ethical keys to the future of nanotechnology'. It will be hard for the sceptics not to appear to have disregarded the need for this quality.

An 'open-ended approach', as characterised by Vogt et al (2007: 332) following Hunt and Mehta, befits the enormity of the task in developing the ethics of nanotechnology. It is no time for parsimony of any sort. As Hans Jonas (1984: 18) has pointed out, the power of new technologies and the tremors they unleash 'call for the utter resources of ethical thought'. Accordingly, nanoethics will be home for all types of ethical thinking of past, present and future. Its distinctive methodology consists in adopting the broadest possible approach that allows for dipping in and out of all the schools in search of an ideal mix or choice. The suggestion is that nanoethics will have to touch on utilitarianism, deontological approaches, virtue and communitarian and other schools of ethics.

This is easier said than done. However, this proposal makes sense when seen in light of the projections of nanoethics as the 'ethics of the largest' (Hunt, 2006: 183). Transatlantic conversations offer a glimpse of the potential of nanoethics to be a meta-level ethics under which the ethics for all technoscience will be subsumed. The European reaction to the US 'NBIC' (Nano, Bio, Info and Cogno) report (Roco & Bainbridge, 2003) was to deconstruct the 'NBIC' version of technological convergence and reconstitute it with an emphatic list of prefixes: *nano, bio, cogno, info, socio, anthropo, philo, geo, eco, urbo, orbo, macro, micro* (European Commission, High Level Expert Group, 2004; Cameron, 2007b: 28).

Nanoethics calls for the identification and resolution of meta-level issues that do not come up as explicitly as they should. It is worth noting that the meta-issue of society's direction is the least discussed point when nanotechnology is introduced, and its deployment is validated (Wejnert, 2004: 328). The encompassing issue, the essence of the meta-ethics that nanotechnology would propel to the foreground, is the question of what kind of society we want to be, and interestingly as to 'who we are' (Cameron, 2007a: 288).

The scope of nanoethics is eminently broad – indeed too broad for a branch of practical ethics. The right question will be whether nanoethics can or should be limited to just a branch of practical ethics. A similar issue is raised with environmental ethics which overlap with nanoethics. The discipline of environmental ethics is thought to be too broad to be a branch of practical ethics since the issues it raises sprawl beyond the spatio-temporal contours of practical ethics. Thus, environmental ethics is held to be more than a branch of practical ethics; rather it is a philosophy in its own right. Philosopher Dale Jamieson (1999: 9) elevates it to a full-fledged new area of philosophy: 'Rather than being a subfield of practical ethics which in turn is a subfield of ethics, environmental philosophy is really a new area of philosophy.' This holds true for nanoethics. The justification for ascribing the status of a new philosophy to nanoethics is evident from the preceding discussion.

Nanoethics as a new philosophy will come up with fresh issues but this does not necessarily entail an *ex nihilo* commencement of a neat ethical discourse. While building upon the wealth of bioethical discourses is a necessary and tremendously helpful approach, it is not necessarily the best approach. Rather, it would be an ideal approach to combat the slack in ethical discourse. After all, how can we be sure that our vision is not tempered by the after-effect of the legacy ethics that was dependent on such phenomena as macro-ontology oblivious to the nanoscale world, scarcity, and greed running the markets? The richness of extant ethical discourse should in no way constrain the tendency to burst apart the ethical tradition and then reconstitute it. There stands the warning that it is 'only after we escape the "deadly embrace" of older illusions can ethics gain a proper footing' (Khushf, 2007b: 191).

NANOETHICS, NANO-DIVIDE, NANO-APARTHEID AND AFRICA

The renowned neuroscientist Baroness Susan Greenfield (2003: 268) warned of a sombre technological future, worse than anything humanity has ever seen:

> *[The Vast Majority] are in danger not only of being disenfranchised from a vastly more comfortable way of life but also of being exploited and abused in ways more sinister, pervasive and cruel than even witnessed by the worst excesses of the colonialist past.*

She is obviously not alone in this dreary forecast. Manuel Castells (1998: 82) has already captured the effect of the 'digital divide' – the uneven deployment of the relatively benign information communication technologies (ICT) – on Africa as 'the de-humanization of Africa'. The plots in many science fiction scenarios revolve around this split, as does the scientific prose examining the scenario (Silver, 1997: 282; Annas, 2005: 51). What is more is that the danger has already been experienced by the developing world even with regard to the deployment of ICT. Philip Emeagwali, the Nigerian IT giant who invented a supercomputer in the 1980s, shares what the Baroness had to say. He warns of a repeat of the horrors of slavery and colonialism comparing ICT with the patently harmless compass that was instrumental in the operations to siphon away the sons and daughters of Africa as slaves. Emeagwali (2007) cites the example of the outsourcing of software programming jobs, in which the actual programmers work for a pittance while those outsourcing the job pocket the real value that the job brings.

Such examples are not isolated cases. This is generally the case for workers at call centres in India and elsewhere. The depressingly familiar arrangement bears the hallmarks of the infamous sweatshops set up in the developing world for export-oriented manufacturing (Pieterse, 2005: 13). Africa is now the preferred destination for outsourcing with the newly branded concept of 'impact sourcing' supposedly aimed at tackling unemployment in Africa. It remains to be seen whether the re-branding of outsourcing as 'impact sourcing' is ethically clean as per Emeagwali's warning.

The advent of nanotechnology means that an exacerbation of the effects of the digital divide is a certainty unless there are ethical, legal and regulatory interventions to mitigate its unwelcome impact. The digital divide will have to be reconsidered in light of the critical role converging technologies play in the socio-economic development of nations across the globe. The 'nano-divide' phenomenon is perceived as a much deeper and wider divide than its precursors and it requires a discourse commensurate with the magnitude of the upheavals that it is poised to trigger.

The term nano-divide, like digital divide, can be interpreted as a neutral descriptive concept and as such, as Pieterse (2005: 18) highlights, its theme is 'unusual because it is ordinary for new technology to spread unevenly'. Technology is introduced to the market with the highest bidder having early access before the price goes down and the technology is made available to more adopters. No matter how it is corrected, the marketplace has always had its winners and losers and a gap between them is a prerequisite for its functioning: '... this is just the way that things are' as Weckert (2007: 60) put it, when summarising this line of argument. Weckert attributes prima facie plausibility to this line of argument, saying that 'there is something almost natural about this'. Seen from this perspective, the term nano-divide does not sufficiently cater for the normative issues arising from an uneven diffusion of technology and its implementation as a means of domination. The nano-divide simply describes what naturally happens and even tends to justify its occurrence: 'if we want improvements we have to put up with the costs'.

The term nano-divide, like the digital divide, has normative connotations. It was the aura of neutrality suggested by the term 'digital divide' that led to the coining of another phrase by General Colin Powell, former US military leader and the first African American to serve as US Secretary of State. He insisted on using 'an even stronger term', 'digital apartheid', in place of the relatively weaker phrase 'digital divide' (Ragnedda, 2017: 13). The same logic has led the European Commission (2004) to adopt 'knowledge apartheid' as a substitute for the nano-divide.

'Nano-apartheid' is a more fitting expression to capture the scenarios described by the term 'nano-divide'. Two decades on after the formal abolition of the social policy that the word 'apartheid' described, 'arguably one of the few political terms known throughout the world' is still in very wide circulation (Giliomee & Schlemmer, 1989: 40). It is often employed to refer to economic inequality irrespective of race and it is increasingly used as a metaphor to capture any discriminatory conduct. Commentators on technology diffusion have also adopted the term apartheid as their preferred metaphor. Castells (1998: 95) used the phrase 'technological apartheid' while discussing the effect of the digital divide on Africa while Pieterse (2005: 18) chose to use the term 'cyber/information apartheid' to refer to those excluded from the information revolution.

The discourses of nano-apartheid, knowledge apartheid and digital apartheid reflect what Mandela (1995) has said of the digital divide: the right to access the technologies is a human right and 'their denial is made an instrument of repression'.

Mandela's evaluation of the denial of ICTs as a means of repression is endorsed in the context of the nano-divide. Commentators troubled by the unsettling speed of the advance of technology and the continuing marginalisation of developing countries find the existing intellectual property regime to be 'an instrument of domination' that retains and strengthens existing divides (Arya quoted in Maclurcan, 2009: 147). Those developing the technology first will exploit their 'prohibitive lead' for economic, political and military domination and hegemony.

Such domination is what the future holds for the vast majority if technology continues to be deployed in the same way as it has been deployed heretofore, i.e., as the rich person's toy. By 'toys for the rich', Freeman Dyson (1997: 197) refers to the literal meaning and to 'the technological conveniences that are available only to a minority of people and make it harder for those excluded to take part in the economic and cultural life of the community'. Nano-apartheid, unlike apartheid proper, will not evolve as a racially oriented discrepancy, nor will it emerge mainly as a divide between countries, but more as a divide between non-geographic constellations (Castells, 1998:

130; BBC, 2007; Summers, 2009). The label often used to refer to the disenfranchised majority is the term 'fourth world'. That has now morphed from a label for countries that are the poorest of the poor to a term signifying 'the new geography of social exclusion' (Castells, 1998: 164). Nano-apartheid is, therefore, a fitting label for the chasm between the technology rich and the rest of the world – the 'fourth world', or to use the phrase by Baroness Greenfield (2003: 268), the vast majority.

It is important to note that Africans remain the major constituents of the fourth world as 32 of the 49 least developed countries are in Africa. The identification of this constellation should be central to debates about nano-apartheid which are set to occupy the central agenda on global development policy in the coming decades.

The fundamental issue that the debate around nano-apartheid needs to address is the imperative to develop and sustain a discourse that would be home for the issue. As technology advances, the interests, agendas and potentials of the technology haves and have-nots will continue to diverge and this lack of a shared interest will prove to be a fundamental challenge. This is a major lesson to be drawn from apartheid proper. It is pointed out that 'apartheid was a form of the politics of difference in that it deliberately prevented the development of social cohesion and hindered the development of a shared moral discourse' (Quoted in Letseka, 2012: 48). Concern about a possible failure to develop a shared moral discourse has been echoed in the debate on nanotechnology and the NBIC convergence. It was this possibility that enraged Leon Fuerth, then Vice President Al Gore's national security advisor, who scolded those techno-elites who intend to exclude the wider public from the debate:

> *These guys talking here act as though the government is not part of their lives. They may wish it weren't, but it is. As we approach the issues they debated here today, they had better believe that those issues will be debated by the whole country. The majority of Americans will not simply sit still while some elite strips off [sic] their personalities and uploads [sic] themselves into their cyberspace paradise. They will have something to say about*

> *that. There will be a vehement debate about that in this country* (quoted in Reynolds, 2003: 180).

Fuerth detected a tendency to block the debate about access and called instead for the opening up and development of the discourse. While the debate Fuerth was yearning for has begun, there is no guarantee that it will lead to the interrogation of the business-as-usual, market-led diffusion of technology and the skewed uneven distribution of the benefits of technology. What would Africans say of themselves considering what Fuerth has to say of Americans? Are the stakes any less significant for Africans?

It may not be the prospect of the elites stripping their minds and uploading themselves into their 'cyberspace paradise' that may trouble Africans today. Even though this should equally concern Africans, the more compelling immediate issue for many is the detrimental effect the nano-revolution may have on Africa. Certainly, it was not Fuerth's troubles that South Africa's former Minister of Science and Technology, Mosibudi Mangena, had in mind when he warned of the impact of nanotechnology on Africa:

> *with the increased investment in nanotechnology research and innovation, most traditional materials in specialised applications will, over time, be replaced by cheaper, functionally rich and stronger nano-materials. It is important to ensure that our natural resources do not become redundant, especially because our economy is still very much dependent on them* (quoted in Miller, 2008: 221).

It is for these same reasons that Africa should be engaging the nanoethics debate in a manner as aggressive as Fuerth's rebuke of the techno-elites. The main immediate concern in Africa is not that of exclusion from the benefits of nanotechnology. Rather it is the disruptive potential of nanotechnology to remove the advantages Africa already has: the advantages of primary products such as rubber, cotton, copper and diamonds will diminish or vanish altogether (ETC Group, 2005). The case of lab-manufactured diamonds that began to shake the global diamond mining and trading industry is a compelling instance and an early taste of things to come.[7] Mined African diamonds will lose their privileged share of the market as purer diamonds 'cultured' in backyard garage workshops take over the market (Maney, 2005). Africa cannot rely on mining its diamonds or on creating a niche market for 'organic diamonds'.

The impact of the nano-revolution on Africa is in no way limited to the loss of markets for certain commodities. The complete marginalisation and the eventual dumping of Africa into irrelevance is not an impossibility unless there is an intervention that would redirect the current trends of uneven technology diffusion. Africa may find itself yearning for the good old days of arm-twisting by old and new multinational corporationss whose interest in Africa's resources is now set to diminish. Africa should raise its voice about the withholding of the benefits of technology because these benefits are capable of eliminating the major, if not all, obstacles to development. This will be a major subject matter of global nanoethics and Africa needs to lead the debate now as it is the continent with the most at stake. At the same time, Africa and other developing countries may find themselves the playground of nanotechnologies, in medicine, industrial production, utilisation of ICTs, management of nano-waste and so on.

NANOETHICS AND SOUTH AFRICA'S 'UNFINISHED BUSINESS'

IBM is one of the pioneer companies in the field of nanotechnology. The scanning tunnelling microscope (STM), the workhorse of

---

7 'Cultured diamond' produced at negligible cost using nanotechnology is purer than organic diamond and with made-to-order qualities. This comes as a nightmare for De Beers, the cartel that dominated the global diamond trade for over a century. It should be more worrying for the countries whose economies have been dependent on the export of diamond, for De Beers is likely to veer towards a strategy to control and exploit the new technology rather than stay in the already strained relationship with exporting countries (Uldrich, 2006: 3; for the behind-the-scene dealings for the distribution of the new artificial diamonds, see Maney, 2005).

nanotechnology, was first made at IBM. One of the most iconic achievements in nanotechnology was also of IBM's making: IBM scientists were able to write the IBM logo using 35 xenon atoms. This was among the most powerful scientific feats of all times. It was this achievement that proved Richard Feynman's claim that there is no principle in physics that speaks against the possibility of moving atoms where we want them and when we want them. IBM have a picture of their atomic logo hanging in their headquarters in Zurich and they have rightly captioned it 'the beginning'.

It may as well be 'the beginning' for their ethical stand as they have not always found the right balance between promoting their business interests and complying with ethical demands. This was particularly true of their operations in South Africa for which they were finally dragged to the dock in US courts under the Alien Tort Claims Act (ATCA). IBM featured among the fifty plus companies that were accused of 'aiding and abetting the crime of apartheid against humanity' in the Khulumani case.[8] The human rights abuses of the apartheid regime were facilitated by computers provided by IBM. IBM ignored the international outcry and the UN Convention on the Suppression and Punishment of the Crime of Apartheid and supplied computers to the apartheid government, which used the computers to commit the crime of apartheid as defined in the convention: 'inhuman acts committed for the purpose of establishing and maintaining domination by one racial group of persons over any other racial group of persons and systematically oppressing them'. The then chairman of IBM, Frank Cary, tried to shrug off the ethical, and evidently legal, obligations claiming that IBM is not responsible for its customers' use of equipment supplied by IBM. He was quoted as saying:

We would not bid any business where we believe that our products are going to be used to abridge human rights. However,

---

8 The Khulumani case is a law suit initiated by the Khulumani Support Group that represents victims of apartheid who were involved in the hearings of the Truth and Reconciliation Commission and other survivors of human rights violations under apartheid. The lawsuit was brought before US courts under the Alien Tort Claims Act of the US (www.khulumani.net).

> *we do not see how IBM or any other computer manufacturer can guarantee that they will not be. The facts of the matter are that we do not and cannot control the actions of our customers ...* (Leonard, 1978: 4).

It was, however, beyond doubt that the apartheid government was using IBM computers for military purposes. In view of the clear international law in place and the US arms embargo restrictions, IBM's action was clearly unethical if not illegal. IBM's excuses did not relieve the company of ethical responsibility. The ethical liability of IBM and the rest of the companies that stand accused off aiding and abetting the apartheid regime has become even more certain with the settlement reached between victims of apartheid repression and the other defendant, General Motors. This company provided the apartheid government with customised vehicles used for repression. While GM stopped short of admitting legal liability, it nonetheless agreed to a $1.5 million settlement – an act that 'amounts to a tacit acknowledgement of responsibility' (Davis, 2012; Khulumani Support Group, 2012).

The IBM case is a reminder that those coming up with new products and innovations need to be aware of their ethical obligations. The Khulumani case remains what some commentators refer to as the 'un-finished business' of the reconciliation process in South Africa (Khulumani Support Group, 2012). Similarly the case against the South African cardiologist, Wouter Basson, known as Doctor Death, who manufactured and supplied the apartheid regime with chemical warfare, is another reminder that ethics of technology in South Africa has a beleaguered history (BBC, 2013). The abuse of technology under apartheid is highly instructive to the debate on what should constitute the subject of nanoethics. The production of nanotechnology products in violation of existing laws will form a clear case of an unethical use of nanotechnology. Even with no clear violation of such laws, the ethical obligations of companies should guide their actions. Nanoethics is expected to provide such guidance.

The role of technology in South Africa's history propels to the fore the question of the neutrality of technology – that technology is neutral and it is only its use that gives rise to moral issues. The argument by the IBM executive quoted above alludes to this theory of the neutrality of technology. However, the axiological neutrality of technology is no longer a defence for developers who are aware of the potential misuse of their technologies.

The ethics of nanotechnology in South Africa has another dimension. South Africa boasts a world-class research infrastructure that has given rise to vibrant nanotechnology research which has, in turn, led to collaborations with advanced countries. South Africa has more research collaborations with North partners than it has with its South counterparts or regional partners (Boshoff, 2010). An important issue for nanoethics in South Africa is whether the rest of Africa can benefit from South Africa's developed nanotechnology infrastructure or whether Africa will lose out in accordance with the imperatives of global competitiveness that is driving nanotechnology innovation around the globe. As South Africa's nanotechnology policy is mainly focused around the promotion of its national interests, the accommodation of pan-African interests in the development of nanotechnology requires a huge commitment on South Africa's side.

South African state policy has explicitly addressed this need to look beyond narrow interests. From the Mandela-Mbeki declaration on African renaissance to the economic strategies of various departments, it has been reiterated that South Africa has its goals set well beyond its borders, especially targeting African development as one of its priorities.

Zooming in on South Africa's nanotechnology initiative, one finds specific expressions of this policy in various initiatives. South Africa's nanotechnology strategy has set as its priority meeting the needs of the historically disadvantaged in South Africa – needs similar to the needs of the rest of Africa. The strategy aims to facilitate the massive problems of poverty in the country. Hence, the focus on nanotechnology to address poverty-related diseases (PRDs), access to safe drinking water and other African problems.

South Africa's nanotechnology endeavours have the potential to narrow the ever-widening gap between the developed world and poor African nations. South Africa's approach may be of significance for other sub-Saharan countries with regard to the need to give voice to underdeveloped and transitional countries in directing global nanotechnology governance towards the promotion of equality (Musee, Foladori & Azoulay, 2012).

South Africa has indeed taken on board its status as the continental leader in nanotechnology and has pledged to direct the technology to tackle African problems. It has several collaborations with African countries on multi-lateral and bilateral bases. The networks under the auspices of the African Union and the New Partnership for Africa's Development (AU-NEPAD) framework provide the basis for Africa-wide science and technology collaborations. The more focused regional arrangements under Southern African Development Community (SADC) provide important templates for the rest of Africa. The lessons to be drawn from SADC activities are valuable in many respects. In addition to continental and regional multi-lateral cooperation, South Africa has several bilateral and trilateral engagements with African nations as part of the implementation of its Focus on Africa Programme (FAP). The programme is funded by the DST and administered by the International Relations and Cooperation (IR&C) directorate, a division of the NRF which is mandated 'to facilitate the internationalisation of science amongst researchers between South Africa and elsewhere in the world' (NRF/RISA, 2011). The mandate with respect to FAP is aimed at facilitating access to regional, continental and international opportunities and resources through cooperative and collaborative engagements. South African nanotechnology policy has the potential to turn nanotechnology from its first-world-focused trajectory to the fight against poverty and to diversify its application in the developing world. Nanoethics makes it imperative that such policy is replicated by the frontier countries leading the development of nanotechnology.

CONCLUSION

This chapter has examined the global debate around nanoethics. It is argued that Africa should enter the global nanoethics debate and play its part in the general agenda-setting to shape the technology. Special emphasis is placed on the term 'nano-apartheid. With the historical resonance it carries in the African context, it is a useful discursive device within nanoethics that could be used to represent the concerns of the continent. It is also important that the intensity of the debate is stepped up to achieve commensurate moral progress. Einstein, reading Haldane's book *Daedalus and Icarus,* shared the thought that 'the progress of science is destined to bring enormous confusion and misery to mankind unless it is accompanied by progress in ethics' (Dyson, 1997: 99). The call for new ethics under the mantra of nanoethics should be considered as a response to the Einstein-Haldane rider that 'ethical progress is the only cure for the damage done by scientific progress' (Dyson, 1997: 200). The field of nanoethics is indispensable if any moral progress commensurate with the power of the technology is to be achieved.

REFERENCES

Allhoff, F & Lin, P. 2006. 'What's so special about nanotechnology and nanoethics?'. *International Journal of Applied Philosophy*, 20(2), pp 179–190.

Annas, GJ. 2005. *American Bioethics: Crossing Human Rights and Health Law Boundaries*. New York: Oxford University Press.

Arras, J. 1997, 'Going down to cases: The revival of casuistry in bioethics'. In Jecker, N, Jonsen AR & Pearlman RA. eds. *Bioethics: An Introduction to History, Methods and Practice*. Sudbury: Jones and Bartlett Publishers, Inc.

BBC. 2007. 'New York hunger levels rising'. Available at: www.news.bbc.co.uk/1/hi/world/americas/7106726.stm (accessed on 2 November 2014).

BBC. 2013. 'South Africa's "Dr Death" Basson found guilty of misconduct'. Available at: www.bbc.com/news/world-africa-25432367 (accessed on 2 November 2014).

Bennett-Woods, D., 2008. *Nanotechnology: Ethics and Society*. USA: CRC Press.

Boshoff, N. 2010. 'South–South research collaboration of countries in the Southern African Development Community (SADC)'. *Scientometrics,* 84(2), pp 481–503.

Cameron, N & Mitchell, ME. eds., 2007. *Nanoscale: Issues and Perspectives for the Nano Century*. New Jersey: John Wiley & Sons.

Cameron, NM. 2007a. 'Toward nanoethics?' In Cameron, N & Mitchell, ME. eds. *Nanoscale: Issues and Perspectives for the Nano Century.* New Jersey:

John Wiley & Sons.

Cameron, NM. 2007b. 'Ethics, policy, and the nanotechnology initiative: The transatlantic debate on "converging technologies"'. In Cameron, N & Mitchell, ME. eds. *Nanoscale: Issues and Perspectives for the Nano Century*. New Jersey: John Wiley & Sons.

Caplan, A. 2008. 'Deciphering nanoethics'. *Chemical and Engineering News*, 86(31 March), p 13.

Castells, M. 1998. *End of Millennium*. UK: Blackwell.

Chesters, G & Welsh, I. 2005. 'Complexity and social movement(s) process and emergence in planetary action systems'. *Theory, Culture & Society*, 22(5), pp 187–211.

Davis, R. 2012. 'General Motors concedes to Khulumani in apartheid reparations case'. *Daily Maverick*, 1 March. Available at: www.dailymaverick.co.za (accessed on 23 December 2016).

Dupuy, JP. 2008. 'Some pitfalls in the philosophical foundations of nanoethics'. *The Journal of Medicine and Philosophy*, 32(3), pp 237–261.

Dyson, F. 1997. *Imagined Worlds – Jerusalem – Harvard Lectures*. Cambridge, MA: Harvard University Press.

Emeagwali, P. 2007. 'Technology is the root of all evil', *Africa News*, 4 December 2007. Available at: www.gamji.com/article6000/NEWS7667.htm (accessed on 10 October 2014).

ETC Group. 2005. *NanoGeoPolitics: ETC Group surveys the political landscape* ETC Group Special Report, July/August 2005 Communiqué No.89. Available at: http://www.etcgroup.org/sites/www.etcgroup.org/files/publication/51/01/com89specialnanopoliticsjul05eng.pdf (accessed on 10 October 2014).

European Commission. 2004. *Towards a European Strategy for Nanotechnology: Communication from the Commission.* Available at: www.cordis.europa.eu/pub/nanotechnology/docs/nano_com_en_new.pdf (accessed on 10 October 2014).

European Commission, High Level Expert Group. 2004. *Foresighting the New Technology Wave: Converging Technologies-Shaping the Future of European Societies.* Available at: www.cordis.europa.eu/pub/foresight/docs/ntw_report_nordmann_final_en.pdf (accessed on 20 November 2014).

Giliomee, H & Schlemmer, L. 1989. *From Apartheid to Nation Building*. Cape Town: Oxford University Press.

Greenfield, S. 2004. *Tomorrow's People: How 21st-century Technology is Changing the Way We Think and Feel.* UK: Penguin Books.

Hunt, G. 2006, 'The global ethics of nanotechnology'. In Hunt, G & Mehta, M. eds. *Nanotechnology: Risk, Ethics and Law*. London: Earthscan.

Hunt, G. 2008, 'Negotiating global priorities for technologies'. *Journal of Industrial Ecology*, 12 (3), pp 275–277.

Hunt, G & Mehta, M. 2006. 'Introduction: The challenge of nanotechnologies'. In Hunt, G & Mehta, M. eds. *Nanotechnology: Risk, Ethics and Law*. London: Earthscan.

Jamieson, D. 1999. 'Singer and the practical ethics movement'. In Jamieson, D. ed. *Singer and His Critiques*. Oxford: Blackwell.

Johnson, DG. 2007. 'Ethics and technology "in the making": An essay on the challenge of nanoethics'. *NanoEthics*, 1(1), pp 21–30.

Jonas, H. 1984. *The Imperative of Responsibility: In Search of an Ethics for the Technological Age*. Chicago: University of Chicago Press.

Kaiser, M. 2006. 'Drawing the boundaries of nanoscience: Rationalizing the concerns?'. *The Journal of Law, Medicine & Ethics*, 34(4), pp 667–674.

Keiper, A. 2007. 'Nanoethics as a discipline?'. *The New Atlantis: A Journal of Technology & Society*, (16), pp 55–67.

Kahn J. 2006. 'Nano's big future: Nanotechnology'. *National Geographic*, 209(6), pp 98–119.

Khulumani Support Group. 2012. 'Breaking news: Bankrupt General Motors agrees to settle in apartheid lawsuit', Available at: www.khulumani.net/khulumani/statements/item/620-breaking-news-bankrupt-general-motors-agrees-to-settle-in-apartheid-lawsuit.html (accessed on 21 November 2014).

Khushf, G. 2006. 'An ethic for enhancing human performance through integrative technologies'. In Bainbridge, WS & Roco, MC, eds. *Managing Nano-Bio-Info-Cogno Innovations: Converging technologies in society*. The Netherlands: Springer, pp 255–278.

Khushf, G. 2007a. 'Open questions in the ethics of convergence'. *Journal of Medicine and Philosophy*, 32(3), pp 299–310.

Khushf, G. 2007b. 'The ethics of NBIC convergence'. *Journal of Medicine and Philosophy*, 32(3), pp 185–196.

Leonard, R. 1978. *Computers in South Africa: A Survey of US Companies*. Africa Fund.

Letseka, M. 2012. 'In defence of ubuntu'. *Studies in Philosophy and Education*, 31(1), pp 47–60.

Lin P & Allhoff F. 2007. 'Nanoscience and nanoethics: Defining the disciplines'. In Allhoff F, Lin P, Moor J & Weckert J. eds. *Nanoethics: The Ethical and Social Implications of Nanotechnology*. Hoboken, New Jersey: Wiley-Interscience, pp 3–16.

Litton, P. 2007. '"Nanoethic"?: What's New?'. *Hastings Center Report*, 37(1), pp 22–25.

MacIntyre, A. 1985. *After Virtue: A Study in Moral Theory*. London: Duckworth.

Maclurcan, DC. 2009. 'Southern roles in global nanotechnology innovation: Perspectives from Thailand and Australia'. *NanoEthics*, 3(2), pp 137–156.

Mandela, N. 1995. 'Address by President Nelson Mandela at the opening ceremony of Telecom 95'. 7th World Telecommunications Forum and Exhibition, 3 October 1995, Geneva. Available at: www.mandela.gov.za/mandela_speeches/1995/951003_telecom.htm (accessed on 21 November 2014).

Maney, K, 2005. 'Man-made diamonds sparkle with potential'. *USA Today*, 6.

McGinn, RE. 2010. 'What's different, ethically, about nanotechnology? Foundational questions and answers'. *Nanoethics*, 4(2), pp 115–128.

Miller G. 2008. 'Contemplating the implications of a nanotechnology "Revolution"'. In Fisher E, Selin C, Wetmore JM. eds. *Presenting Futures: The Yearbook of Nanotechnology in Society, vol 1*. Dordrecht: Springer, pp 215–225.Moor, JH. 2005. 'Why we need better ethics for emerging technologies'. *Ethics and Information Technology*, 7(3), pp 111–119.

Moore, C. 2006. 'Killing the Bayh-Dole Act's golden goose'. *Tulane Journal of Technology and Intellectual Property*, 8, p 151.

Moriarty, P. 2008. 'Reclaiming academia from post-academia'. *Nature Nanotechnology*, 3(2), pp 60–62.

Musee, N, Foladori, G & Azoulay, D. 2012. 'Social and environmental implications of nanotechnology development in Africa'. *CSIR (Nanotechnology Environmental Impacts Research Group, South Africa)/ReLANS (Latin American Nanotechnology and Society Network)/IPEN (International POPs Elimination Network)*. Available at: www.ipen.org/pdfs/nano_booklet_sept_5.pdf (accessed on 21 November 14).

NRF/RISA. 2011. *International Relations and Cooperation: Focus on Africa Programme* (Pocket Guide). Pretoria: NRF/RISA (National Research Foundation/Research and Innovation Support and Advancement).

O'Neill, O. 2002, *Autonomy and Trust in Bioethics*. Cambridge, UK: Cambridge University Press.

Parens, E, Johnston, J & Moses, J. 2008. 'Do we need "synthetic bioethics"'. *Science*, 321(5895), p 1449.

Pieterse, JN. 2005. 'Digital capitalism and development: The unbearable lightness of ICT4D'. In Lovink, G & Zehle, S, eds. *Incommunicado Reader: Information Technology for Everybody Else*. Amsterdam: Institute of Network Cultures.

Ragnedda, M. 2017. *The Third Digital Divide: A Weberian Approach to Digital Inequalities*. London: Taylor & Francis.

Reynolds, GH. 2003. 'Nanotechnology and regulatory policy: Three futures'. *Harvard Journal of Law & Technology*, 17(1), pp 179–225.

Roco, MC & Bainbridge, WS. eds. 2002. *Converging Technologies for Improving Human Performance: Nanotechnology, Biotechnology, Information Technology and the Cognitive Science*. Arlington, VA: National Science Foundation.

Sandler, R. 2007. 'Nanotechnology and social context'. *Bulletin of Science, Technology & Society*, 27(6), pp 446–454.

Schwandt, TR. 1990. 'Paths to inquiry in the social disciplines: Scientific, constructivist, and critical theory methodologies'. In Guba, EG. ed. *The Paradigm Dialog*. London: Sage Publications, pp 258–276.
Silver, LM. 1998. *Remaking Eden: How Genetic Engineering and Cloning Will Transform the American Family*. Princeton: Avon Books.
Somerville, M. 2004. *The Ethical Canary: Science, Society and the Human Spirit*. Montreal: McGill Queen's University Press.
Summers, L. 2009. 'Address on international development and the global economy', Granoff Forum. Available at: www.sas.upenn.edu/home/news/granoff09.pdf (accessed on 21 November 2014)
Ten Have, HAMJ. 2007. 'Introduction: UNESCO, ethics and emerging technologies'. *Nanotechnologies, Ethics and Politics.* Paris: UNESCO Publishing.
Uldrich, J. 2006. *Investing in Nanotechnology: Think Small. Win Big*. USA: Simon and Schuster.
Vogt, T, Baird, D & Robinson, C. 2007. 'Opportunities in the "post-academic" world. *Nature Nanotechnology*, 2(6), pp 329–332.
Wamboga-Mugirya, P. 2008. 'Museveni urges science journalists to plug knowledge gap'. Available at: www.scidev.net/en/news/museveni-urges-science-journalists-to-plug-knowled.html (accessed on 21 November 2014).
Weckert, J. 2007. 'An approach to nanoethics'. In Hodge, GA, Bowman, DM & Ludlow, K. eds. *New Global Frontiers in Regulation: The Age of Nanotechnology*. Cheltenham, UK: Edward Elgar Publishing, pp 49–66.
Wejnert, J. 2004. 'Regulatory mechanisms for molecular nanotechnology'. *Jurimetrics*, 44(3), pp 323–350.

FOUR

Diseases of poverty

Nanomedicine research in South Africa

Thomas S Woodson

INTRODUCTION

Since 2000, there has been increasing pressure to address global public health concerns like high child mortality rates and a lack of essential medicines for the poor. In 2000, a variety of factors came together to push new global health initiatives. First, the United Nations (UN) launched the Millennium Development Goals (MDGs) to tackle global poverty. Of the eight MDGs, three of them targeted health issues like reducing child mortality, improving maternal health, and combating HIV/AIDS and malaria (United Nations, 2014). In addition, a number of celebrity activists like Bono, Angelina Jolie and Bill Clinton started advocating for large initiatives to end global poverty (Cohen, 2006). Finally, large non-profit organisations, like the Bill & Melinda Gates Foundation, and government aid agencies, like the Swedish International Development Cooperation (SIDA), created public–private partnerships (PPPs) to develop new medicines to eradicate diseases of poverty (Cohen, 2006).

Over the same period of time, nanotechnology research and development (R&D) rose in prominence. In 2000, the USA launched the National Nanotechnology Initiative and by 2004 over 60 countries had nanotechnology programmes (Maclurcan, 2010). Nanotechnology became the exciting emerging technology that would transform a wide range of sectors, ranging from energy generation to sports equipment. In the healthcare sector, many argued that nanotechnologies would revolutionise medicine by creating new drug delivery systems, sensors and prosthetics (Wagner et al, 2006).

South Africa was a part of these trends and the country started a nanotechnology initiative with healthcare as one of its six focus areas (Claassens & Motuku, 2006). The country spent about R170 million (US$26 million) on its nanotechnology initiative and opened nanotechnology research centres across the country. Within the national nanotechnology strategy, the South African Department of Science and Technology (DST) explicitly stated that it wanted the nanotechnology initiative to address poverty and inequality in South Africa. However, after 10 years of investment, have their efforts been successful? Is nanomedicine addressing illnesses that have a high disease burden in South Africa? Secondly, what is the role of PPPs and product development partnerships (PDPs) for medicine development in South Africa? Globally, PPPs and PDPs are active in developing medicines and vaccines for diseases of poverty (Moran et al, 2010). Are these organisations partnering with nanomedicine researchers in South Africa to use the latest technology to create medicines for the poor? This study examines whether PPPs are a part of the nanotechnology revolution in South Africa and if they can serve a useful purpose to bring this technology to the market.

LITERATURE REVIEW

Nanotechnology

The idea that scientists would manipulate matter at the nanoscale in order to build new devices and products arose around the 1960s. Richard Feynman, a physicist, imagined a future where scientists would write an encyclopaedia on the head of a pin or build tiny motors only 100 atoms across (Feynman, 1959). Eventually, scientists developed microscopes that allowed them to observe and manipulate atoms. Scientists saw that there was a new world of science at the nanoscale that could revolutionise materials and, as a result, scientists began advocating for more funding for nanotechnology (Gallo, 2009). Finally in 2000, the USA started the National Nanotechnology Initiative and the nanotechnology race began in earnest. By 2004, over 60 countries had nanotechnology initiatives (Maclurcan, 2010) and in that year alone, national governments gave over US$3.7 billion towards nanotechnology R&D (Roco, 2005).

Nanotechnology is used in a variety of fields ranging from energy storage to consumer sports equipment, but from its inception, scholars envisioned the potential of nanotechnology for health applications. In Feynman's seminal speech, he foresaw a day when there would be a 'mechanical surgeon inside the blood vessel and it goes into the heart and "looks" around ... It finds out which valve is the faulty one and takes a little knife and slices it out' (Feynman, 1959). Scientists have not developed these types of devices yet, but they are making targeted drug delivery systems that will attack only the sick cells and leave the rest of the body intact. These new drugs will also allow treatment regimens to be shorter and less complicated (Mamo et al, 2010). They are also designing personalised biological implants that will end organ rejection and are developing new diagnostics tools that allow doctors to detect dozens of diseases with only one drop of blood (Wagner et al, 2006).

Many national nanotechnology initiatives, including those in South Africa, prioritised nanomedicine in their strategies. In 2005, South Africa developed its first nanotechnology strategy with initial funding of R170 million (US$26 million) to target six strategic areas (Claassens & Motuku, 2006). Though this is small by global standards, it was a first step in starting a nanotechnology programme.[9] Within the nanomedicine area, the South African Department of Science

9 In comparison, in 2004 the USA, China and South Korea spent US$986 million, US$200 million and US$300 million, respectively, on nanotechnology R&D (Roco, 2005).

and Technology (DST) said it would pursue nanomedicine in order to 'improve drug delivery systems, including traditional medicine through packaging medicine for ailments such as TB, HIV/AIDS and malaria in nano-capsules'[10] (Department of Science and Technology South Africa, 2005). The DST's commitment to nanomedicine was reaffirmed in 2012 with the Nanoscience and Nanotechnology Plan. In the plan, the South African government expressed that the country needs to develop new technology like point-of-care diagnostics. These diagnostics tools have three important uses: they would speed up drug detection and prevent the spread of disease by letting people know they are sick before they infect others. Moreover, the new diagnostics would decrease false test results. Third, the diagnostic tools would be easier to use so that the country would not need highly trained nurses and doctors to administer the tests. Easy and reliable diagnostics are especially important in poor, rural areas where it is hard to find specialised tests or highly skilled medical personnel (Department of Science and Technology, 2012).

Since the DST started the nanotechnology initiative, the country has had some successes. South Africa has produced several high-profile nanomedicine technologies and programmes. At the Council for Scientific and Industrial Research (CSIR), Hilda Swai is developing nano-encapsulated medicines for TB and HIV (Makoni, 2010). Swai's work has garnered a lot of attention from the international community and she has appeared in a variety of articles about nanotechnology in Africa (Makoni, 2010). In another programme, CSIR hosted a Pan African Nanomedicine Summer School in nanomedicine in 2012. The seven-day summer school had about 130 participants and it featured over fifty lectures on topics ranging from antiviral therapeutics to cell targeting and polymers (Council for Scientific and Industrial Research, 2013).

However, many believe that South Africa has underperformed in

10 HIV/AIDS caused the most premature deaths in South Africa. 48% of years of life lost in South Africa in 2010 were due to HIV/AIDS. Tuberculosis was the fifth leading cause of life lost in South Africa for 2010 and this disease caused 3.3% of premature deaths (Institute for Health Metrics and Evaluation, 2014).

nanotechnology. In 2007, Pouris found that from 2000–2005 South Africa underperformed in nanotechnology research compared to countries like Brazil, India and South Korea. Only 0.8% of South African research publications in Thomson Reuters ISI database were related to nanotechnology, while Brazil and India had 2.1% and 3.6% of their respective research publications related to nanotechnology (Pouris, 2007). In another study, Claasens and Motuku discuss a variety of aspects of South Africa's nanotechnology efforts and they find that South Africa is in the 'middle ground' for nanotechnology development. South Africa is just entering the nanotechnology field and very few companies promote themselves (Claassens & Motuku, 2006). Though these studies provide useful insights into South Africa's nanotechnology programmes, it is necessary to re-evaluate nanotechnology in South Africa since it formally began 10 years ago.

Public–private partnerships

The term public–private partnership (PPP) first arose 40 years ago (Bovaird, 2004), although there is a longer history of governments partnering with private sector entities. In the 16th and 17th centuries, privateer shipping companies, like the Dutch-East India Company, performed many government services like coining money, administering laws and marshalling armies (Wettenhall, 2005). Later during World War II, companies played a large role in developing materials and products for the war effort, and after the war, they were instrumental in international development (Hounshell, 1992). Today, politicians regularly launch PPPs in order to find novel solutions to problems.

A commonly used definition of PPPs is 'working arrangements based on a mutual commitment (over and above that implied in any contract) between a public sector organization with any organization outside of the public sector' (Bovaird, 2004). Bovaird's definition is broad and can incorporate many types of organisations and interactions. For example, according to Bovaird's definition, a PPP does not have to consist of a government organisation and a private sector company, but rather, it can also be between a government and a non-profit organisation. A crucial part of Bovaird's definition is that the relationship has to be 'over and above that implied in any contract' (Bovaird, 2004). Bovaird does not give detailed specifications about what is necessary to be 'over and above' but this factor makes PPPs different from other organisational alliances and contracts.

Over the past 40 years, there has been an increasing emphasis on developing PPPs. PPPs are seen as a way to handle increasingly complex problems that need input from both the public and private sector. For example, many infrastructure projects use a public–private partnership model to build and service toll roads, bridges and ports (Miraftab, 2004). Another important feature of PPPs is that they overcome market barriers that a single actor can face. For example, one organisation may be hesitant to do research on potentially useful yet challenging technology because the cost of failure is too high for them. However, by partnering with other firms, the organisation spreads it risk and this allows it to invest in multiple technology platforms.

Similar to Bovaird's definition, the South African government defines PPPs as 'a contract between a public sector institution/ municipality and a private party, in which the private party assumes substantial financial, technical and operational risk in the design, financing, building and operation of a project' (GTAC, 2015). The South Africa government established an interdepartmental task force to develop policies to govern PPPs in 2000. The government realised that PPPs could play a crucial role in providing new infrastructure and mitigating service delivery backlogs (Fourie, 2008). Over the past 15 years, South Africans have set up a variety of PPPs and currently the South African National Treasury oversees PPPs in 17 sectors including health, energy, water, tourism, waste and business development (National Treasury of South Africa, 2015).

This chapter focuses on a subset of PPPs, called product development partnerships (PDPs), which do healthcare research and development (R&D) and are at the centre of research into diseases of poverty. These organisations do a variety of things like developing new medicines and diagnostics, educating communities about safe health practices, and developing better supply chain and pricing structures to improve access to medicines for individuals in poor countries.

Chataway and others define PDPs as 'social technology innovations designed to develop and distribute physical technologies in the shape of new products and drugs' (Chataway et al, 2009). A common organisational structure of health PDPs is that a large multinational non-profit organisation, like the Bill & Melinda Gates Foundation, will partner with a government aid organisation, like USAID, to form a PDP that targets a specific disease (Woodson, 2014). Previous studies identified about 30 health PPPs and PDPs that engage in a variety of activities like developing malaria medicines and creating new medical diagnostics (Moran et al, 2010; Woodson, 2014).

METHODS

For this study, I use both qualitative and quantitative methods to describe nanomedicine and diseases of poverty R&D in South Africa. First, I used bibliometric techniques to characterise nanomedicine R&D in South Africa. Then I conducted phone interviews of South African PPPs and nanomedicine scholars to understand whether South African PPPs are actively involved in nanomedicine R&D. The interviews also give insights into the progress of nanomedicine R&D in South Africa and how the country can improve its focus to have the biggest impact on South Africans.

Bibliometrics is a common method to measure research output and the innovation environment of a country or region. Bibliometricians have studied myriad topics including Southern Africa research collaboration (Boshoff, 2009), South African research output (Kahn, 2011), nanomedicine (Woodson, 2012), and tropical medicines (Falagas, Karavasiou, & Bliziotis, 2006). For this analysis, I use data from the online, subscription-based Web of Science (WoS). WoS is one of the largest repositories of scholarly publications and it contains over 14 000 academic journals from both natural and social sciences (Clarivate, 2018). Moreover, it is commonly used in bibliometric analyses (UNESCO, 2005).

To begin my analysis, I used nanotechnology databases created by the Center for Nanotechnology and Society at Arizona State

University and the Georgia Institute of Technology Programme in Science, Technology and Innovation Policy (Arora et al, 2012). These databases contain all the nanotechnology articles in WoS from 1990–2013. I also used the WoS internal search engine to find articles authored by South African scholars.

Table 4.1: Keywords for nanomedicine search

Alzheimer	dental	neuron
anaemia	disease	orthopedic
antibiotic	dopamine	pharma
antitumour	drug	physiological
blood	HIV	skin
brain	insulin	therapeutic
cancer	liver	tissue
cholesterol	measles	vaccine
clinic	medicine	wound

Source: Author generated

Within that database, I searched for nanomedicine articles using a keyword strategy that I developed. Table 4.1 shows the keywords I used to find nanomedicine articles. I developed these keywords through a series of steps. First, I generated a large list (about 150 terms) of medical-related keywords from prominent nanomedicine papers. Then, I removed keywords that did not generate unique nanomedicine articles. Third, I tested that at least 70% of the articles generated by a particular keyword related to nanomedicine. This ensured that I was analysing articles related to nanomedicine and not general biotechnology. Next, I asked a nanotechnology expert to verify that the keyword strategy was sufficiently accurate. After several iterations, I created a nanotechnology database that contains nanomedicine publications from 2000–2012. The nanotechnology expert reviewed a random sample of the articles in the database and concluded that 87% of the articles were about nanomedicine. The misclassified articles were mostly about biological nanotechnology R&D and the nanomedicine expert did not consider these articles specific enough to nanomedicine.

Though bibliometrics is a standard method to study research output, it has limitations. Many of the databases that scholars use for bibliometrics are biased toward English-language journals and journals in Western countries (UNESCO, 2005). This means that the research in non-English speaking and non-Western countries is underrepresented. Another limitation of bibliometrics is that it does not account for R&D that is not published or patented. There are a lot of innovations that remain trade secrets or the R&D of these innovations occurs outside of sectors that publish and patent their discoveries. This is especially a problem when considering innovations by poor and marginalised communities. Many innovations in these context are 'below the radar innovations' and will not appear in bibliometric studies (Kaplinsky et al, 2009).

In addition to bibliometric data from WoS, I used global health statistics from the Institute for Health Metrics and Evaluation (IHME). IHME collects disease burden statistics for 291 diseases for every country and region in the world (Murray et al, 2012).

After collecting the bibliometric data, I conducted semi-structured interviews of South African PPPs managers, nanotechnology scientists and government officials. I identified the interview participants by first contacting all the known health PPPs in South African. I also contacted prominent South African nanomedicine scholars that I identified from the bibliometric data. Finally, I used so-called snowball techniques – asking the interviewees to suggest other participants – to generate further interview contacts with PPPs and nanomedicine scholars. I conducted six interviews related to South African nanomedicine and 10 interviews related to global nanomedicine and PPP research. Table 4.2 lists South African health PPPs that I found by reviewing the healthcare literature and talking with healthcare providers in South Africa.

Table 4.2 South African PPPs

SAAVI	Desmond Tutu HIV Foundation
IAVI South Africa	Qhakaza Mbokodo Research Clinic
SATVI	Madibeng Centre for Research
AERAS	Ndlovu
EDCTP-Africa Office	IVI
TB Alliance	National Institute for Communicable Diseases
IDRI	PATH South Africa
IPM	BioVacInstitute

Source: Author generated

RESULTS

From 2000–2012, South Africa's total research output has steadily grown. In 2000, South African scientists published about 5 100 articles in WoS-indexed journals and 1 350 of them were related to the healthcare/biomedical fields. By 2012, South African scientists published about 14 000 articles in WoS-indexed journals and 3 900 of the articles were related to healthcare. Similarly over this time, nanotechnology research started to gain prominence. In 2000, South Africa published almost no articles related to nanomedicine but over the next 10 years there was an exponential rise in the nanomedicine publications (see Figure 4.1). Overall between 2000 and 2012 South African researchers published 214 nanomedicine articles.

Figure 4.1: South African nanomedicine publications by year

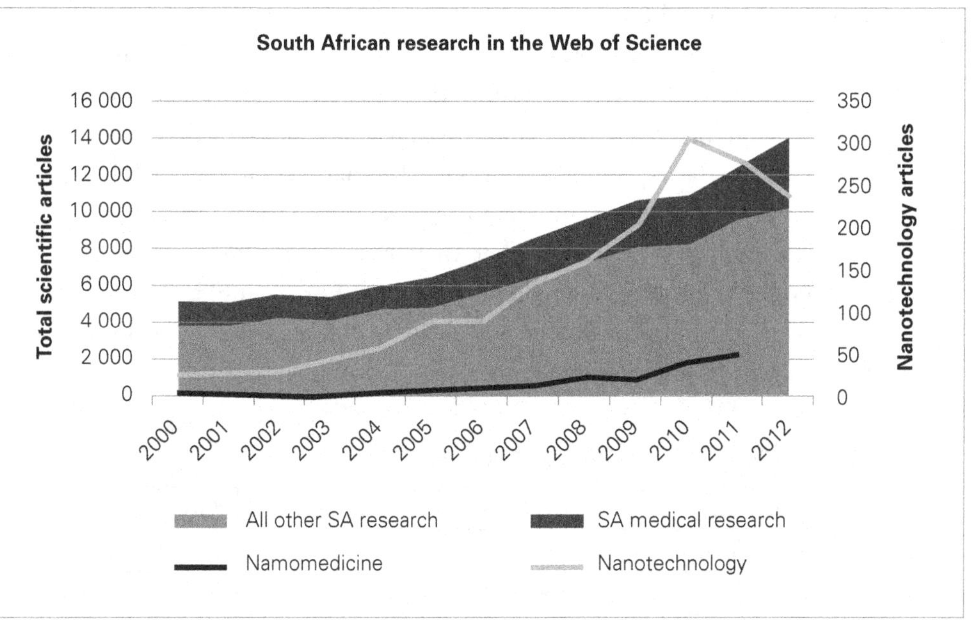

Most of South Africa's published scientific research was done by large universities, like Rhodes University and the University of the Witwatersrand, and national research organisations, like CSIR and MINTEK. This trend holds for nanomedicine research as well. However, the traditionally strong research universities, like the University of Cape Town, the University of the Witwatersrand, the University of Pretoria and Stellenbosch University, do not produce the most nanomedicine research. Rather, the University of Cape Town, which has the most scientific publications (Kahn, 2011), ranks sixth in nanomedicine research. On the other hand, Rhodes University has the most nanomedicine publications, but it ranks sixth in overall scientific output in South Africa (Kahn, 2011). Similarly, CSIR has the fourth most nanomedicine publications, but it is ranked 15th in overall research output (Kahn, 2011).

There are several possible reasons for this difference in research output between nanomedicine research and overall scientific output. First, South Africa established many centres of excellence that serve as a base for nanotechnology research. At these centres, the government installed the expensive equipment necessary for nanotechnology research and then the government allowed researchers from around the country to use these tools for their projects (Department of Science and Technology, 2012). As a result, the centres of excellence, like CSIR, have an advantage over other organisations because they have easy access to the equipment necessary for nanotechnology R&D. Another reason that less renowned organisations are top performers in nanotechnology research is that they have a few key scientists that conduct a lot of nanomedicine research. For example, Rhodes University is the home of Professor Tebello Nyokong who is a leading scholar is sensors and nanomedicine.

Table 4.3 Top South African organisations doing nanomedicine research

	Affiliation	# Records
1	Rhodes University	43
2	University of the Witwatersrand	40
3	University of Pretoria	29
4	CSIR	26
5	University of KwaZulu-Natal	26
6	University of Cape Town	20
7	University of Stellenbosch	15
8	North West University	13
9	University of the Western Cape	12
10	MINTEK	5
11	University of Johannesburg	5
12	AstraZeneca R&D Sodertalje	3
13	University of the Free State	3
14	University of Limpopo	3
15	University of the Orange Free State	3

When I analysed the diseases that the South African nanomedicine publications address (columns 1 and 2 of Table 4.4), it emerged that the top disease studied is tuberculosis, followed by HIV/AIDS and malaria. This is a different pattern to global nanotechnology R&D that focuses on diseases with a high disease burden in wealthy nations like breast cancer, psychological diseases and Alzheimer's (Woodson, 2012). These results suggest that South Africa's nanotechnology initiative successfully directs its research to be more pro-poor (Claassens & Motuku, 2006).

Column 4 of Table 4.4 lists the global 2010 disability adjusted life years (DALYs) of different diseases. The most serious disease on the list is malaria followed by HIV/AIDS. Malaria caused 83 million DALYs in 2010 and HIV/AIDS caused 82 million DALYs in 2010. Column 5 shows the global DALY as a per cent of total DALY and column 6 gives the ratio of the percentage of nanomedicine articles to the percentage of global DALYs. This ratio represents whether South African scholars are over- or understudying a particular disease based on its global DALY. If the ratio is one, then South African scholars are studying a particular disease in proportion to its severity. If the ratio is much greater than one, then researchers are researching the disease in higher proportions than its DALY. From the analysis, we can see that some diseases, like HIV/AIDS and malaria, receive proportional attention from South African scholars. However, Parkinson's disease receives significantly more attention compared to the severity of the disease.

Columns 7–9 repeat the analysis with the South African DALY. This gives a different perspective on South African research foci. In this analysis, several diseases are over-studied compared to the DALY in South Africa. The analysis again shows that South Africans do more nanomedicine research on Parkinson's disease compared to its DALY. The ratio of the percentage of Parkinson's disease publications to the percentage of South African DALYs is 52.24. This means that the proportion of nanotechnology articles on Parkinson's disease compared to Parkinson's disease DALY is 52 times greater than what would be expected if the percentage of publications matched the disease burden. Parkinson's disease is 2% of nanomedicine research, but it only 0.04% of the South African DALY. In addition to Parkinson's disease, South African scholars study malaria, hepatitis B and Alzheimer's in greater proportion than their DALY. On the other hand, HIV/AIDS is severely understudied. HIV/AIDS is 40% of South Africa DALY but it represents only 4% of South African nanomedicine research.

Table 4.4: Nanomedicine research by disease area. DALY statistics come from the World Health Organisation and the Institute for Health Metrics and Evaluation

Disease	# of articles	% articles	Global 2010 DALYs (millions)	% Global DALY	Ratio: % articles to % global DALY	South Africa 2010 DALYs (millions)	% DALY SA	Ratio: % articles to % DALY SA	Clinical trials in South Africa 2000–2014
Tuberculosis	17	0.08	49	0.020	4.02	0.87	0.029	2.70	125
HIV/AIDS	9	0.04	82	0.033	1.29	11.92	0.400	0.11	343
Malaria	7	0.03	83	0.033	0.99	0.03	0.001	36.39	6
Alzheimer/ dementias	6	0.03	11	0.005	6.18	0.05	0.002	18.23	29
Parkinson's disease	4	0.02	2	0.001	24.39	0.01	0.0004	52.24	18
Breast cancer	3	0.01	12	0.005	2.92	0.07	0.002	5.93	91
Hepatitis B	3	0.01	5	0.002	7.51	0.02	0.001	27.31	19
Asthma	2	0.01	22	0.009	1.04	0.22	0.007	1.25	119
Periodontal disease	2	0.01	5	0.002	4.32	0.03	0.001	10.22	1
Typhoid fever	2	0.01	12	0.005	1.91	0.17	0.006	1.59	0
Diabetes mellitus	1	.005	47	0.019	0.25	0.59	0.02	0.24	226
Edentulism	1	.005	5	0.002	2.53	0.03	0.001	4.91	0

This chart illuminates trends in nanomedicine research, but it has limitations. The lack of nanotechnology R&D for diseases with high DALY may be due to the fact that nanotechnology is not the appropriate technology to address those diseases. This study cannot determine whether or not nanotechnology is appropriate for certain diseases, but only that it is not being studied in relation to the severity of the disease. Also, South Africa has so few nanomedicine publications that a few publications on a particular disease will have a big impact on the relative importance of a research area. For example, there is only one nanomedicine publication on edentulism, yet it appears that scientists are focusing too much on this disease compared to its DALY.

Below is a correlation matrix that shows the relationship between the number of nanomedicine articles per disease area, world DALY and South African DALY. There is not a very large correlation between the variables. The number of articles has a higher correlation with world DALY (0.54) compared to South African DALYs (0.34). However, I cannot make any statistical inferences about the relationship between the variables because there are too few observations.

Table 4.5: Correlation between nanomedicine articles, world DALY and South Africa DALY

	Articles	World DALY	SA DALY
Articles	1.00	0.54	0.34
World DALY	0.54	1.00	0.60
SA DALY	0.34	0.60	1.00

INTERVIEW RESULTS

The interview data provide an in-depth look at nanomedicine R&D. The bibliometric analysis characterises the research effort, but the interviews help explain the reason particular patterns emerge. One finding to emerge from the analysis is that South Africa has a strong presence of international PPPs compared to other African countries. Many of the largest health PPPs like the International Aids Vaccine

Initiative (IAVI) and PATH have offices in the country. South Africa is an important partner for PPPs for several reasons. First, many of the PPPs are developing new medicines for diseases of poverty, like tuberculosis and HIV/AIDS, and they are using South Africa as a site for clinical trials. South Africa is an ideal place to host clinical trials because it has a large population suffering from these diseases and it has high-quality research facilities that can run rigorous clinical trials. Clinical trials are long, complicated processes and they have to be done in stable environments with plenty of resources. One manager from a USA-based PPP says that 'South Africa is actually incredibly well developed in TB research. They really emerge as a leader. They sort of have the most advanced research labs and folks there ... So South Africa is definitely big for us'. According ClinicalTrials.gov, Africa has had about 4 460 clinical trials since 2000 and South Africa hosts about half (2 008) of them. Of those trials, 343 of them were on HIV/AIDS and 125 of them were on tuberculosis. Table 4.3 show the clinical trials per disease area.

Another question I asked the PPP managers was why they focused on certain diseases. Almost all of the PPPs are working on HIV/AIDS or tuberculosis. Both of these illnesses are major causes of death and disability in South Africa. HIV/AIDS is 40% of South African DALYs and tuberculosis is 30% of South African DALYs. PPPs cite these high incidence rates as the main reason they are working on TB and HIV/AIDS. One PPP manager in South Africa says that the 'TB problem in South Africa is completely running out of hand. We are one of the highest countries with prevalence and I think we have 870 new case annum per 100 000 inhabitants.' Another PPP manager says that TB is a large problem because 'with the global travel that everyone does it is now a rising problem in Europe. It's in the US, it's all over the world now and it's really making a strong comeback because in order to be really cured of TB you need to be on medicines for 6 months and that's a very long period of time for people to adhere. They might feel better in two months and then they'll stop their medicine and then the resistant bugs will start to grow. So drug-resistant TB is a really big issue. I think that is the focus of many PPPs beyond HIV'.

Another South African PPP manager says that her organisation focuses on HIV because there is a 'massive gap in access to treatment, and in access to care, and South Africa lacks the total capacity to manage it at this stage'.

In addition to treating TB and HIV/AIDS, South African PPPs focus on building local capacity. These PPPs want to train local scientists and engineers to do the tests, run the experiments and make the medicines instead of watching American and European researchers do so. The PPPs believe that building local capacity is central to South Africa's development. However, it is often hard to build capacity in poor and rural places. One manager expressed frustration about the difficulties of building capacity by saying that 'the most important resource is the human resource and it is not easy to get top people drawn to a rural area'.

A third goal of the PPPs that became apparent in the interviews is to bridge the gap between the public sector and the private sector. This was made clear by several organisations like the South African Technology Innovation Agency (TIA). One person described the TIA as an important agency because it 'is more towards the end of the value chain so in terms of PPPs, they are very important because they have to bridge the gap between scientific visibility and towards market application and relevance to the economy'.

Nanotechnology with PPPs

The next major question relates to PPPs' work with nanotechnology. Overall, PPPs do not work with nanotechnology. Only a few of the PPP managers that I interviewed know about nanotechnology and none of them actually do research on it. However, none of the South African PPP managers were against nanotechnology. One manager said that he liked the idea of nanomedicine but that he is currently too busy to pursue that line of research. He says, 'It's just that we have so many things on our plate and we would rather take one thing at a time. So nanotechnology is not something that we have gone deep into it, but we would be open to engagement with anybody that would be in that space.' Another South African PPP manager expressed similar sentiments, saying that the organisation would be interested in a nanotechnology collaboration.

Challenges of PPPs

Finally, I explored some of the challenges PPPs face. One challenge PPPs have is that they are constantly raising money to fund their R&D. There is limited money available for organisations researching diseases of poverty and since the global recession in 2008, there are even fewer donors able to give to health PPPs. One PPP manager said that donors are quick to stop supporting them which has consequences for their ability to do R&D and provide services: 'I always sarcastically say donors are the first to demand sustainability but are always the first ones to leave after a few years which makes it difficult. That also explains our variety of donors…'

Even the successful PPPs have trouble accessing enough money to maintain a high quality of care and research. One PPP manager said that they have too many patients and individuals seeking treatment. She says, '…you've got a certain budget but you attract so many patients and so many clients that finally you have to say no to more clients because your budget becomes constrained.'

A second challenge for health PPPs in South Africa is that government regulations slow down their operations. One PPP had to build a new treatment centre to sufficiently separate the regular patients from those patients participating in medical trials. This was because the clinic could not mix the two types of patients without violating the law. The regulations protect patients who are not willing to participant in clinical trials, but they add burdens to the PPPs. Another PPP manager complained that regulations hinder their innovation. When asked to explain further which types of government regulations constrain innovation, she said, 'I can't cut it into little bits because it's the whole package, the whole package needs to be looked at and streamlined and cheapened.' Obviously, it's impossible to eliminate all regulations but the government can examine policies with a view to protecting its population and encouraging innovations in PPPs.

Finally, a reoccurring comment made by the PPP managers and government officials is that they want South Africa to move beyond being a great clinical trial site. In a 2012 *Nature* interview, the director of South African Medical Research Council emphasised that being known as a clinical trial site has its disadvantages. He says, 'The problem is that we now need to go beyond being a site for international study to being scientists that lead and initiate these studies. It's one thing to collect data, and another to dream up the idea, design the protocol and do the study' (Maxmen, 2012). South African officials want to develop the human capital of the country so that local researchers can design clinical trials and projects instead of relying on foreign expertise.

CONCLUSION

South Africa's nanomedicine efforts have grown from almost nothing in 2000 to about 50 articles a year in 2011. Over that time, South Africa made a strategic decision to focus on diseases of poverty, and as a result, it has strengths in nanotechnology R&D for tuberculosis and HIV/AIDS. These diseases are a large part of South Africa's disease burden, and hence, the country's R&D strategy fits its need. However, in addition to diseases of poverty, South African scholars have published nanotechnology research on diseases like Alzheimer's and Parkinson's disease. Based solely on their current disease burden, South Africa's investment in these types of diseases may seem unnecessary. However, as the country gets richer, its disease burden will shift from communicable diseases to non-communicable diseases, like cancer and mental illness and, therefore, the country must develop medicines for those diseases as well.

The role that PPPs play in nanomedicine development is minimal. Though these organisations are important to the overall health system, they are not interacting with nanotechnology. The PPPs say that their resources are already strained and that they do not have the time or the ability to pursue nanotechnology. However, they are willing to work with other organisations interested in nanotechnology. Therefore, to improve South Africa's nanomedicine, the country must link its nanotechnology research efforts to the PPPs that are operating in poor and rural communities. The government can do this by providing additional funding for nanomedicine scholars to work with PPPs, and by working with PPPs and scientists to develop policies that help innovation and protect the public safety.

ACKNOWLEDGEMENTS

I would like to thank Shih-Hsin Chen, the Georgia Institute of Technology: Technology Policy Assessment Center, and the anonymous reviewer for their help on this paper. This research was funded by Center for Nanotechnology in Society at Arizona State University with an NSF grant #0937591.

REFERENCES

Arora, SK, Porter, AL, Youtie, J & Shapira, P. 2012. 'Capturing new developments in an emerging technology: An updated search strategy for identifying nanotechnology research outputs'. *Scientometrics*, 95(1), pp 351–370.

Boshoff, N. 2009. 'South–South research collaboration of countries in the Southern African Development Community (SADC)'. *Scientometrics*, 84(2), pp 481–503. doi:10.1007/s11192-009-0120-0.

Bovaird, T. 2004. 'Public–private partnerships: From contested concepts to prevalent practice'. *International Review of Administrative Sciences*, 70(2), pp 199–215. doi:10.1177/0020852304044250.

Chataway, J, Hanlin, R, Muraguri, L & Wamae, W. 2009. 'PDPs as social technology innovators in global health: Operating above and below the radar'. In Penea, O. ed. *Innovating for the Health of All*. Geneva, Switzerland: Global Forum for Health Research.

Claassens, C & Motuku, M. 2006. 'Nanoscience and nanotechnology research and development in South Africa. *Nanotechnology Law & Business*, 3(2), pp 217. Available at: www.scholar.google.com/scholar?hl=en&btnG=Search&q=intitle:Nanoscience+and+Nanotechnology+Research+and+Development+in+South+Africa#0 (accessed in July 2016).

Clarivate, 2018. Available at: www.clarivate.com/products/web-of-science/web-science-form/web-science-core-collection/ (accessed on 17 January 2018).

Cohen, J. 2006. 'The new world of global health'. *Science*, 311(January), pp 162–167.

Council for Scientific and Industrial Research. 2013. '1st Pan-African Summer

School in Nanomedicine'. Available at: www.csir.co.za/msm/nano_summerschool2012/ (accessed on 5 May 2015).

Department of Science and Technology. 2012. 'Nanoscience and nanotechnology: 10-year research plan'. Available at: www.dst.gov.za/images/Research__Plan__Final_small.pdf (accessed in July 2016).

Department of Science and Technology South Africa. 2005. 'The National Nanotechnology Strategy'. Available at: www.esastap.org.za/download/natstrat_nano_2006.pdf (accessed in July 2016).

Falagas, ME, Karavasiou, AI & Bliziotis, IA. 2006. 'A bibliometric analysis of global trends of research productivity in tropical medicine'. *Acta Tropica*, 99(2-3), pp 155–9. doi:10.1016/j.actatropica.2006.07.011.

Feynman, RP. 1959. 'Plenty of room at the bottom'. *APS Annual Meeting*. Available at: www.pa.msu.edu/~yang/RFeynman_plentySpace.pdf (accessed in July 2016).

Fourie, D. 2008. 'The contribution of public private partnerships to economic growth and human capital development: A South African experience'. *Journal of Public Administration*, 43(4), pp 559–570.

Gallo, J. 2009. 'The discursive and operational foundations of the National Nanotechnology Initiative in the history of the National Science Foundation'. *Perspectives on Science*, 17(2), pp 174–211.

GTAC. 2015. 'Transaction Advisory Services (PPP)'. Available at: www.perma.cc/54EF-7YEY (accessed in July 2016).

Hounshell, DA. 1992. 'Du Pont and the management of large-scale research and development'. In Galison, P & Hevly, B, eds. *Big Science: The Growth of Large-Scale Research*. Stanford: Stanford University Press.

Institute for Health Metrics and Evaluation. 2014. GBD PROFILE: SOUTH AFRICA. Seattle.

Kahn, M. 2011. 'A bibliometric analysis of South Africa's scientific outputs: Some trends and implications'. *South African Journal of Science*, 107(1/2). doi:10.4102/sajs.v107i1/2.406.

Kaplinsky, R, Chataway, J, Clark, N, Hanlin, R, Kale, D, Muraguri, L, Papaioannou, T, Robbins, P & Wamae, W. 2009. 'Below the radar: What does innovation in emerging economies have to offer other low-income economies?'. *International Journal of Technology Management & Sustainable Development*, 8(3), pp 177–197. doi:10.1386/ijtm8.3.177/1.

Leydesdorff, L, Carley, S & Rafols, I. 2013. 'Global maps of science based on the new Web-of-Science categories'. *Scientometrics*, 94(2), pp 589–593. doi:10.1007/s11192-012-0784-8.

Maclurcan, DC. 2010. 'Nanotechnology and the hope for a more equitable world: A mixed methods study'. Sydney: University of Technology.

Makoni, M. 2010. 'Case study: South Africa uses nanotech against TB'. Available at: www.scidev.net/global/health/feature/case-study-south-africa-uses-nanotech-against-tb-1.html (accessed in July 2016).

Mamo, T, Moseman, EA, Kolishetti, N, Salvador-Morales, C, Shi, J, Kuritzkes, DR, Langer, R, Von Andrian, U & Farokhzad, OC. 2010. 'Emerging nanotechnology approaches for HIV/AIDS treatment and prevention'. *Nanomedicine*, 5(2), pp 269–285. doi:10.2217/nnm.10.1.

Maxmen, A. 2012. 'A slim chance for South African medical research'. *Nature*, (June), pp 1–4.

Miraftab, F. 2004. 'Public–private partnerships: The Trojan Horse of neoliberal development?' *Journal of Planning Education and Research*, 24(1), pp 89–101. doi:10.1177/0739456X04267173.

Moran, M, Guzman, J, Ropars, AL & Illmer, A. 2010. 'The role of product development partnerships in research and development for neglected diseases'. *International Health*, 2(2), pp 114–122. doi:10.1016/j.inhe.2010.04.002.

Murray, CJ, Vos, T, Lozano, R, Naghavi, M, Flaxman, AD, Michaud, C, Ezzati, M, Shibuya, K, Salomon, JA, Abdalla, S & Aboyans, V. 2012. 'Disability-adjusted life years (DALYs) for 291 diseases and injuries in 21 regions, 1990–2010: A systematic analysis for the Global Burden of Disease Study 2010'. *The Lancet*, 380(9859), pp 2197–2223.

Pouris, A. 2007. 'Nanoscale research in South Africa: A mapping exercise based on scientometrics'. *Scientometrics*, 70(3), pp 541–553. doi:10.1007/s11192-007-0301-7.

Roco, MC. 2005. 'International perspective on government nanotechnology funding in 2005'. *Journal of Nanoparticle Research*, 7(6), pp 707–712. Available at: www.springerlink.com/index/7T5L75V66X52N885.pdf (accessed in July 2016).

UNESCO. 2005. *What do Bibliometric Indicators Tell Us About World Scientific Output?* (Vol. 2). Montreal: UNESCO. Available at: www.csiic.ca/PDF/UIS_bulletin_sept2005_EN.pdf (accessed in July 2016).

United Nations. 2014. 'United Nations Millennium Development Goals'. Available at: www.un.org/millenniumgoals/ (accessed on 8 December 2014).

Wagner, V, Dullaart, A, Bock, AK & Zweck, A. 2006. 'The emerging nanomedicine landscape'. *Nature Biotechnology*, 24(10), pp 1211–1218. doi:10.1038/nbt1006-1211

Wettenhall, R. 2005. 'The public–private interface: Surveying the history'. In Hodge, G & Greve, C. eds. *The Challenge of Public–Private Partnerships: Learning from International Experience*. Cheltenham, UK: Edward Edgar Publishing.

Woodson, TS. 2012. 'Research inequality in nanomedicine'. *Journal of Business Chemistry*, 9(3), pp 133–146.

Woodson, TS. 2014. 'Emerging technologies for the poor: How nanomedicine and public-private partnerships are used to address diseases of poverty'. PhD thesis, Georgia Institute of Technology, Atlanta, Georgia.

FIVE

Building a bio-economy in South Africa

Lessons from biotechnology innovation networks in Taiwan

SHIH-HSIN CHEN & THOMAS S WOODSON

INTRODUCTION

THE BIOTECHNOLOGY INDUSTRY is an intensive knowledge-based sector and many economies, both big and small, want to have thriving biotechnology sectors (Balaguer et al, 2008). Countries have heavily invested in both creating a strong bioscience base and supporting small- and medium-sized biotechnology enterprises, using a range of policy tools (Senker et al, 2007). In these countries, particular attention has been paid to the creation of new start-up firms, and there has been a number of recent policy initiatives designed to stimulate commercialisation within the biotechnology industry (Senker et al, 2007).

South Africa launched its bio-economy strategy in early 2014, with a focus on growing the economy and using biotechnology to achieve a positive socio-economic impact. At the other end of world, the 2014 scientific board meeting of Biotechnology Committee (BTC) of Taiwan concluded that Taiwan would launch the Bio-economy Blueprint (of Taiwan) by 2016 with the aim of developing the technology.

Considering the similar ambitions of building up the biotechnology industry, it is interesting to compare South Africa's and Taiwan's bio-economies. South Africa and Taiwan are such different countries and are almost polar opposites in terms of history, culture and institutions, yet they have a common goal of developing a bio-economy. This chapter briefly describes the evolution of the bio-economies in Taiwan and South Africa, and how cultural and institutional factors shaped two widely differing biotechnology strategies.

The chapter begins by reviewing the literature on bio-economy, network theory and knowledge production. This will help us better understand innovation development in South Africa and Taiwan in order to form a basis for further comparing the development of innovation networks while building up the bio-economy.

LITERATURE REVIEW

Building innovation networks in the biotechnology sector

Since Enriquez-Cabot proposed the concept of bio-economy in 1998, this concept has gradually emerged in policy documents worldwide, especially in the 2010s. 'Bio-economy refers to all economic activity derived from scientific and research activity focused on biotechnology industrial process' (Enriquez-Cabot, 1998: 925–926). Biotechnology is the application of the principles of engineering and biological science to create new products from raw materials of biological origin (e.g., vaccines or food) (Verma et al 2011). All over the world, countries and regions have invested in the bio-economy as a means to increase economic growth and prosperity (Staffas, Gustavsson & McCormick, 2013). The OECD wrote a policy White Paper, namely 'The bio-economy to 2030: Designing a policy agenda' in the early 2000s with the expectation of developing biotechnology to result in an emerging 'bio-economy' (Organisation for Economic Co-operation and Development & OECD International Futures Programme, 2009).

The policy agenda of 'The bio-economy to 2030' anticipated that the development of biotechnology would contribute to a significant share of global economic output (Organisation for Economic Co-operation and Development & OECD International Futures Programme, 2009).

In the USA, the Obama administration built the US bio-economy as a means of retaining the dominance of US research capacity. The White House was focused on enhancing technology transfer and public–private partnerships (PPP) (White House, 2012: 1 & 5). The EU, Germany, Finland, Sweden and Australia also adopted some type of bio-economy strategy in the late 2000s and early 2010s (Staffas, Gustavsson & McCormicj, 2013). Like most country strategies, they all depend on unique definitions for bio-economy and use slightly different tactics for improving their bio-economies. Some of the plans emphasise sustainability, while other strategies predominantly discuss innovation. Despite the diversity of strategies, many of them discuss collaboration, networks and working across sectors to build their strategies (Staffas, Gustavsson & McCormicj, 2013).

Networks of scientists, engineers and researchers are often considered a key factor in innovation (Freeman, 1991; Giuliani, 2011). However, until the recent development of computational and visualisation tools, such as those deployed in social network analysis (Powell, 2013), it was difficult to study the impacts of networks on innovation. Social network analysis was originally developed as a sociology tool to analyse the relationships between the actors in the network (Freeman, 2011). Since the 1970s, scholars, such as Granovetter (1973), Burt (2001) and Coleman (1988; 1994) expanded the application of systematic social network analysis and linked their observations on network structures with social theories, such as the theory of social capital. These have subsequently been widely adopted in innovation studies in recent years (Asheim, 2011).

Network theory, in general, refers to several different kinds of ideas which concern 'the proposed processes and mechanisms that relate network properties to the outcomes of interests' (Marin & Wellman, 2011: 40 in Scott & Carrington, 2011). At its earlier stages, network theory was mainly applied for studying information flow between actors, commonly referred to as nodes. According to Powell

(2013), 'Networks are where knowledge resides and action transpires. Networks have been largely invisible, but recent developments in mapping nets have led to an outpouring of work on network topology and dynamics.' The concepts of information networks have been adopted in the mainstream innovation literature. As Freeman pointed out, studies shifted focus to the 'vital importance of external information networks and of collaboration with users during the development of new products and processes' (1991: 499). In the last two decades, the literature has focused on the analysis of network position and structure, and on the associated effects of innovation networks (Giuliani, 2011). Consequently, research has been concerned with the determinants or consequences of networks as well as actors' positions, and roles within networks have become more prominent in innovation studies.

Alongside networks, knowledge production has long played a crucial role in the modern economy (Leydesdorff & Zawdie, 2010). In the 1990s, the literature moved from the linear model of R&D into the interactive mode. This shift was marked by analyses of society's role in shaping the knowledge base and knowledge production in academia (Pavitt, 1998) and the emergence of the innovation systems' literature. The OECD suggested, given the importance of knowledge networks, that 'the firm-level innovation study needs to be developed to better characterise innovation processes and interactions among firms and a range of institutional actors in the economy' (1996: 43).

Although the literature mostly adopts 'knowledge networks' and 'knowledge economy' without clearly defining them (Stiglitz, 1999; Brinkley, 2006; Dolfsma & Soete, 2006; Vallas & Kleinman, 2008), a number of scholars have identified the characteristics of the knowledge economy. For instance, Powell and Snellman (2004) defined a key component of a knowledge economy to be a greater reliance on intellectual capabilities than on physical inputs or natural resources. Based on their study of the confluence of academic and commercial biotechnology innovation in the US, Vallas and Kleinman (2008) suggest that the knowledge economy has begun to emerge across previously distinct institutional domains. Leydesdorff (2006) developed the triple helix model which allows for interaction effects among domains and specific synergies among functions and institutions of science and technology, government and industry in the knowledge-based economy.

In innovation studies, the terminology 'innovation intermediaries' (Hargadon, 1998; Burt, 2004; Howells, 2006) has been used to examine firms, individuals and third parties that enable innovation by transferring knowledge (Meyer & Kearnes, 2013). Recent literature also demonstrates the diverse possibilities of various actors being knowledge brokers in the innovation system: for example, discussion of the feasibilities of firms as knowledge brokers (Hargadon & Sutton, 1997; Hargadon, 1998); suggestions about the role private technology-generating sectors may play (Den Hertog & Bilderbeek, 2000; Kaghan, 2001; Fisher & Atkinson-Grosjean, 2002; Tether & Tajar, 2008); analysis of the role of venture capital managers in enhancing knowledge transfer (Zook, 2004); and descriptions of the role of university-based technology transfer offices in the knowledge transformation process (Lerner, 2004). It has been demonstrated in the literature that different types of organisations, including firms, venture capitalists and university-based technology transfer offices, can function as brokers or intermediaries in an innovation system. These brokers typically facilitate access to and mastery of technology with the ultimate goal of enhancing innovation. Consequently, Yusuf (2008) concludes that specialised and institutional intermediaries can facilitate knowledge transfer and strengthen the effectiveness of the knowledge networks between industry and the leading universities (which could be considered as general purpose intermediaries).

DATA AND METHODS

For this chapter, we first conducted an in-depth literature review of the bio-economy in South Africa and Taiwan. We present the history of the different bio-economy sectors and how the governments encouraged it. Then we did a bibliometric analysis of biotechnology publications in South Africa and Taiwan and interviewed several biotechnology experts to better understand the challenges in the bio-economies of South Africa and Taiwan.

For the bibliometrics, we started by studying collaboration in South Africa and Taiwan over the past three decades using scientometrics mapping techniques developed by Leydesdorff and others (Leydesdorff & Persson, 2010; Leydesdorff et al, 2012). To analyse the network developments over the past three decades, we analysed the data from 1983 to 2013. We chose these years because at the time of the analysis, 2013 was the most recent year with complete bibliometric data. We then analysed the bio-economy 15 years and 30 years prior to 2013. By observing several different time points, we can see how the networks changed over time. For the network analysis, we searched the ISI Web of Science database research area (SU) for Taiwanese and South African publications from 1983, 1998, and 2013 using the keyword 'biotechnology'.

After downloading the data, we produced geographical collaboration network maps (like Figure 5.1) using the methods of Leydesdorff et al (Leydesdorff & Persson, 2010; Leydesdorff et al, 2012).[11]

Next we produced subject category maps (like Figure 5.3) using tools for mapping PubMed Data. PubMed, like Web of Science, is an online repository for research. However, PubMed focuses on biomedicine research (National Institutes of Health, 2013).[12] The two key search terms we used to find articles were:

(Taiwan [Affiliation]) AND biotech*

(South Africa [Affiliation]) AND biotech*

We searched for articles in May 2014, and found 4 201 Taiwanese articles and 1 410 South African articles.

Finally, we conducted a small number (six) of selective interviews with the key actors of firms, research organisations, non-profit organisations and intermediaries in South Africa, including the institutional intermediaries and NGOs involved in the development of the biotechnology sector. Most of the interview data were from elite informants involved in technology transfer in the biotechnology sector in South Africa. Interviews were semi-structured and ranged between 45 and 120 minutes in length and were digitally recorded.

---

11 The instructions can be found online at www.leydesdorff.net/maps/.

12 The detailed steps of this set of mapping can be found online at www.leydesdorff.net/pubmed/.

The interview questions ask about the roles these agents play in the innovation and development process of developing the biotechnology sector in South Africa.

ANALYSIS AND DISCUSSION

Innovation and development in South Africa

South Africa has a long history. In the pre-colonial period, pastorial communities occupied most of the landscape, and it has been established that many of these settlements engaged in sophisticated agricultural production, mining and international trade. Europeans first came to South Africa in the 15th century. Over the next 400 years, more Europeans (mainly Dutch and British) came to South Africa to set up trading posts, exploit agricultural opportunities and mine the country's rich natural resources (South African Communication and Information System, 2011). The Europeans brought the racist policies that segregated the country and created two separate economies, at the same time as they introduced modern infrastructure and economic activities tailored along European lines. These policies of racial segregation and exclusion weakened the innovation system of South Africa, isolating the country from the rest of the world and forcing the country to look inward (Bruhn & Gallego, 2012).

Despite its isolation and human injustice, the 'white' South African economy initially thrived. This part of South African society successfully developed industries like military, mining and energy complexes. But South Africa's economic success under apartheid was short-lived. In the 1970s the country was affected by the oil embargo and the growing strain of apartheid, and over the next 10 years it became obvious that the system was unstable. For example, under apartheid, science and technology programmes did not receive enough funding, and research and development (R&D) activities were confined to the white community to serve its interests. The Council for Scientific and Industrial Research (CSIR), which was South Africa's premier research institution, lost much of its prestige, relevance and innovative spirit under the apartheid system (Scerri, 2009). Moreover, there was little private investment in R&D because the country did not offer tax incentives to companies to conduct R&D, and as a result, private R&D was very low. Mining was the only sector that had significant private R&D and innovation.

In 1994, South Africa held democratic elections (Scerri, 2009; South African Communication and Information System, 2011), and made the transition to an inclusive, non-racial and democractic system. After the transition in 1994, South Africa reformed almost every aspect of society. The country reworked the education system, improved healthcare, provided housing and sanitation to the majority of the population, and opened its borders to trade. In 1996, the government issued a policy white paper called the 'National Research and Development Strategy' on the science and technology development that set up the country's innovation system 'based on the core principles of partnerships, co-ordination, problem-solving and multi-disciplinary knowledge production' (Kruss & Lorentzen, 2009). The white paper attempted to embed S&T strategies within a larger drive towards achieving a winning National System of Innovation in South Africa. The White Paper on Science and Technology (1996) encouraged stakeholders to forge collaborative partnerships in order to integrate resources from engineering, the natural sciences, the health sciences, environmental sciences and social sciences for problem-solving in a multi-disciplinary manner. Even though health science was mentioned as one of the technology domains in the white paper, biotechnology was not addressed as a key discipline. The white paper encouraged both applied and basic research and it encouraged technology transfer.

South Africa's innovation strategy was aggressive and it has helped to advance R&D in the country. General expenditure on research and development (GERD) in South Africa has grown steadily since 1993. Between 1993 and 2004 GERD went from 0.6% to 0.87% (Kaplan, 2008). Moreover, the country's private GERD is 56.3% of the total GERD and about 15% of R&D funding comes from foreign sources. These levels of GERD are high for developing countries (Kaplan, 2008). South Africa also has a good tertiary education system. The country attracts the brightest students from Africa and approximately 30% of South African PhD students are from other African nations

(Department of Science and Technology South Africa, 2011). Overall, the OECD believes that South Africa's innovation system has achieved extensive reforms and shown tremendous growth since 1993 (Kaplan, 2008).

However, South Africa still faces many challenges. In 2008, 22% of people lived below the poverty line. This is a significant improvement from the 38% poverty rate in 2000 (World Bank, 2011), but there are still about 11 million people who live below the poverty line. Also, South Africa does not have enough scientists and researchers to become a top performer in R&D. It only has 364 researchers per million people. Though this is a lot of research capacity for developing countries, it is low for more developed ones. For example, in 2010 the USA, China and Brazil respectively had 3 838; 890; and 710 researchers for one million people. India, however, had 137 researchers per million people (World Bank, 2014b).

In addition, South Africa's overall percentage of global publications has fallen sharply from its peak in 1987. In 1987, South Africa had 0.7% of global publications but in 2003 they only had 0.48% of global publications. While South Africa lost its global publication presence, South Korea, Brazil, Taiwan and India have increased their global share of publications (Kaplan, 2008). South Africa also declined in the world rankings in the knowledge economy index calculated by the World Bank (Kaplan, 2008). This, it seems, is more reflective of the rate at which the other countries have advanced in R&D activities, rather than stagnation within South Africa.

In 2001, the Department of Science and Technology (DST) developed the National Biotechnology Strategy to initiate the development of biotechnology and associated products and services that would address the science-based innovation needs in the health, industrial and agricultural sectors of the economy (2001). Biotechnology was officially documented in the policy report as a key technology domain to articulate innovation and development in multiple technology application domains. Following the launch of the National Biotechnology Strategy (2001), several regional biotechnology innovation centres, such as LIFElab, BioPad, CapeBiotech, and PlantBio and the knowledge transfer intermediaries such as Technology

Innovation Agency (TIA) were built for promoting the development of biotechnology to strengthen the agriculture and health, as well as the industry and environment sectors. DST attempted to coordinate the government fund to attract private investment to encourage joint funding across the public–private sector. The TIA was anticipated to play the mediating role in biotechnology communications and marketing but it has been facing several systemic barriers, such as financial challenges (Interview, WS450269).

More than one decade later, in early 2014, the South African Bio-economy Strategy positioned bio-innovation as essential to the achievement of the government's industrial and social development goals in South Africa. DST was positioned as the lead agent to coordinate the work with other departments. TIA plays the key mediating role in the commercialisation of manufacturing, and biotechnology commercialisation.

Innovation and development in Taiwan

Taiwan consists of a large island and a series of small isles. The islands were colonised by the Dutch in the 17th century, followed by a long history of being part of mainland China. In 1895, Taiwan, along with Penghu (another small island), was ceded to Japan; and for six decades Taiwan was a colony of Japan until the end of World War II. Since 1945, Taiwan seceded from mainland China to establish the Republic of China (ROC), though the People's Republic of China considers it part of ROC in the context of the globally acknowledged 'one China policy'.

Taiwan's population is about 23 million people. Its 2013 GDP was US$926.4 billion and the GDP per capita was US$39 600 (Central Intelligence Agency, 2016). It has a small economy, but is ranked as a middle-to-high income economy. In 2013, Taiwan spent about US$23 billion on R&D. This was 2.4% of GDP (Central Intelligence Agency, 2016).

In 2011, the Taiwanese researchers published about 26 650 SCI papers, which meant it ranked as the 16th most productive region/country around the world. Compared with the amount of SCI publications that Taiwanese scholars published in 2002 (about 11 436 articles), the number of articles grew by 2.33 times between 2002 and 2011. In terms of a five-year relative impact (RI) of citation rate, 2007–2011, the RI of Taiwan is 0.76, which is lower than the RI of Singapore (1.19) and Japan (1.02), but resembles the RI of South Korea (0.78) and mainland China (0.76) (STPI, 2012: 96–97)

In the 1980s, the Statute for Upgrading Industries was important for transforming the overall industrial structure in Taiwan, from a manufacturing-based industry into a knowledge-intensive economy. Furthermore, the amendment of the University Act in 2005 significantly stimulated the academia–industry collaboration in the second half of the 2000s. The Foundation Law for Technology Development, enacted in 1999, allowed the universities to become patent assignees and be able to exercise patent ownership.

The emergence of the policy instruments in the 2000s has jointly contributed to the development of the biotechnology industry and enhanced the establishment of academia–industry collaborations. To upgrade the industrial structure of Taiwan, the leadership launched the Science and Technology Development Plan in 1982 and the statute for upgrading industries (SUI) became effective in 1991. The statute was set to expire at the end of 2010. Before 2010 it was not clear whether there would be any replacement afterwards. However, the 'Statutes for Industrial Innovation', promulgated on 12 May 2010, follows up on the SUI to enhance innovation development in agricultural, industrial and service sectors. During this process, the government preferred to implement policies through institutional intermediaries, in particular, the development centre of biotechnology.

The first time that biotechnology was included in an official policy document of Taiwan was in 1982 in the Science and Technology Development Plan (Bianchi et al, 2011; BPIPO, 2011; Wang, 2011). In 1978, energy, material, information and automation were selected as four key science and technology areas. Later, in 1982, biotechnology, optical technology, food technology and hepatitis prevention were further taken into account. Altogether, these eight issues were designated as the 'Eight Key Technologies'. In addition, in 1982, for the first time, the government decided to invest direct support for the research and development of vaccine technology, which was the first official record of government intervention dedicated to biopharmaceutical innovation in Taiwan. The government established a vaccine manufacturing firm, namely Life Guard Pharmaceuticals (LGP) and the Development Centre for Biotechnology (DCB) in 1984, to take on the mission of developing vaccine technology and promoting the biotechnology industry in Taiwan. The Development Centre for Biotechnology (DCB) was established to promote the biotechnology industry, starting with agricultural biotechnology. The policies and interventions of the government also targeted a broadly defined area of biotechnology.

In August 1995, the first policy plan specifically for the biotechnology industry was the 'Promotion Plan for the Biotechnology Industry'. The aim of the plan was 'developing the biotechnology industry, reinforcing biotechnology research and development, and enhancing the international competitiveness of the biotechnology industry in Taiwan'. According to the Promotion Plan for the Biotechnology Industry (1995), the government in Taiwan aggressively supported and promoted the biotechnology industry. The support included encouraging the private sector to invest in biotechnology, promoting international collaboration and investment, constructing government-owned science parks and improving biotechnology-related regulations. Since the early 2000s, DCB has shifted R&D focus onto pharmaceuticals, herbal medicine and biopharmaceuticals. Compared with other industries, the biotechnology industry stands out as one of the few sectors which have received capital investment directly from the government through the National Development Fund (NDF).

Since the arrival of the policy recommendation made by the BioTaiwan Committee in 2005 and 2006, the Taiwanese government has started to shift its attention and concentrate policy to focus on biopharmaceuticals, agriculture biotechnology and medical devices (DCB, 2008b). The official establishment of the National Development Fund (NDF) in 2006, the promulgation of the Biotech and New Pharmaceutical Development Act (BNPDA) in 2007, and the launch of the Taiwan Diamond Action Plan for Biotech Takeoff in 2009 actively strengthened the policy framework in order to promote the biotechnology industry in Taiwan.

The other two main policy instruments arrived even later, in the second half of the 2000s: Biotech and the New Pharmaceutical Development Act (BNPDA) in 2007 and Taiwan Biotech Take-off Diamond Action Plan 2009. Considering the US White House launched the Bio-economy Blue Print in 2012 with the aim of strengthening the PPP (public–private partnership) in the biotechnology sector, in 2014, the Taiwanese BTC biannual meeting proposed an ambitious plan to launch the Taiwan Bio-economy Industry Development Program in 2016 (BOST, 2016) .

SCIENTIFIC RESEARCH IN BIOTECHNOLOGY

The different strategies of Taiwan and South Africa are evident in the type of research that each country produces. Figure 5.1 indicates that the South African biotechnology research mainly focuses on agricultural biotechnology, in particular in the field of animal disease (fish disease, swine disease, building animal disease model and so on). Taiwan's biotechnology research, Figure 5.2, mainly focuses on bioreactor[13] and lipopolysaccharides[14]. These differences make sense in light of the expertise of each country and the focus of their bio-economy strategies. In South Africa, food stuffs, vegetable products and animal product industries make up 9.3% of total exports while chemical products make up 6%. Package medicaments are only 0.34% of exports (Simoes & Hidalgo, 2011). Overall, agriculture makes up 3% of South Africa's GDP. On the other hand, in Taiwan, agriculture is 1.7% of GDP (World Bank, 2014a).

13 A bioreactor is any device or system within which a biological reaction takes place.

14 Lipopolysaccharides are molecules found on the surface of cells and are implicated in strong immune responses in animals.

Figure 5.1: Research focus of South Africa's biotechnology publications

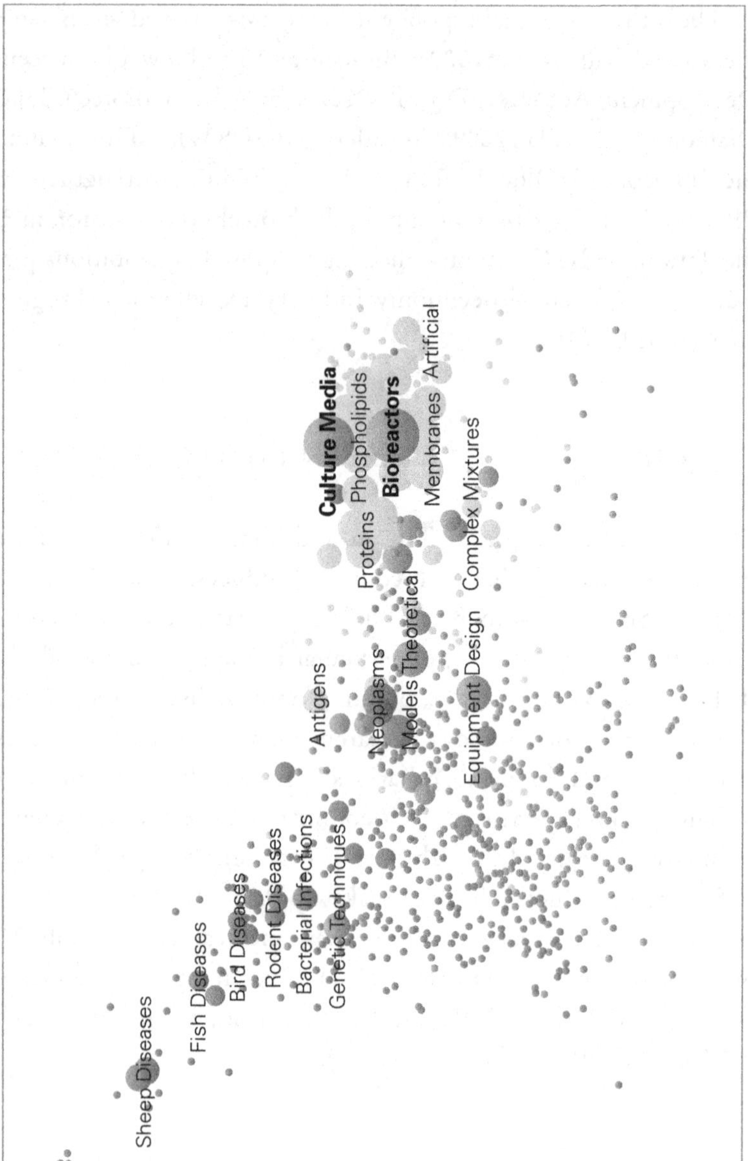

Figure 5.2: Research focus of Taiwan biotechnology publication

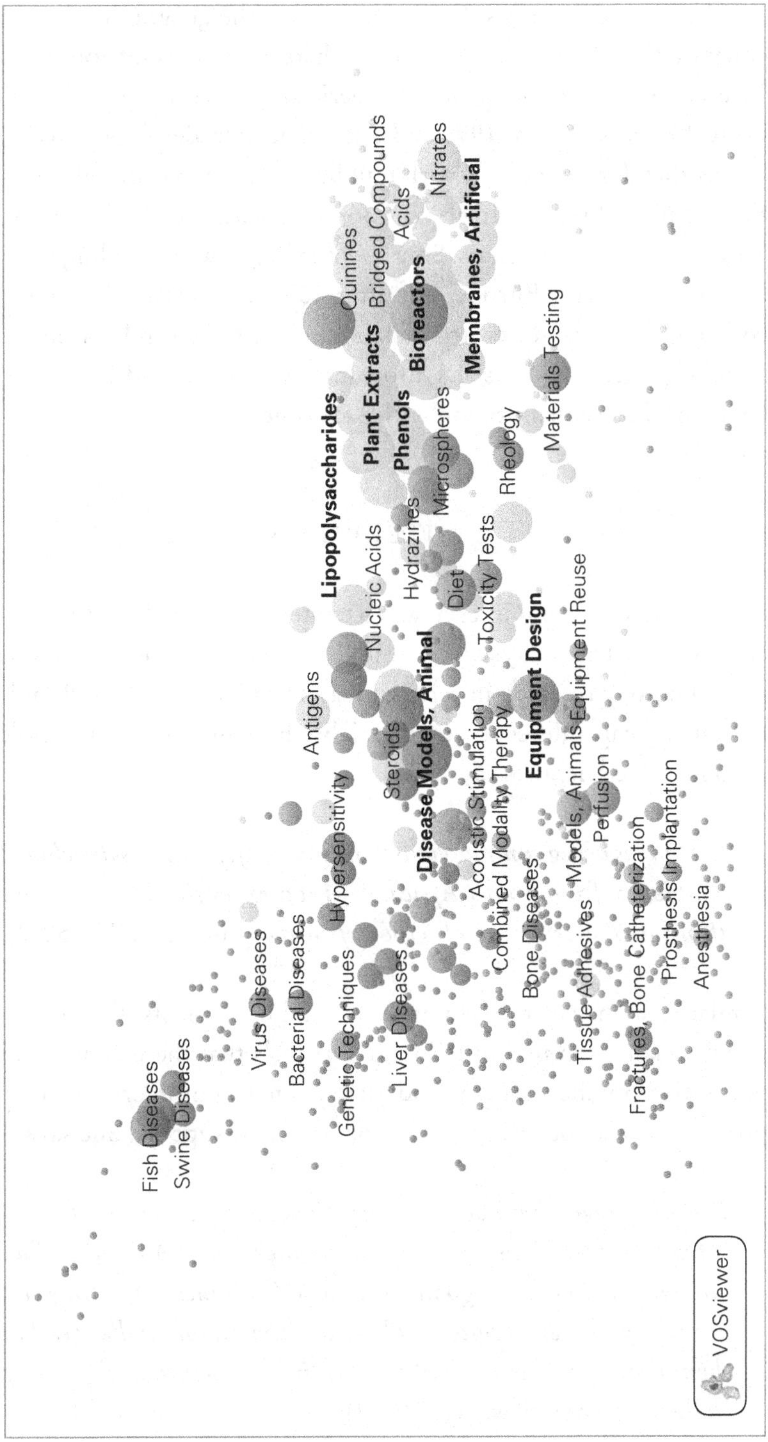

The next set of maps, Figure 5.3, shows the geographical research collaborations by the academic researchers of Taiwan and South Africa. In each region, we map out the periods of research collaborations every 15 years – 1983, 1998 and 2013. The map developed over time shows that Taiwan started with building research collaboration in the field of biotechnology from collaboration with North American and West European researchers. South Africa began with building research collaboration with Europe, but after 2000 has expanded its research collaboration networks to collaborate with researchers in Latin America, Africa and Asia. Unfortunately, Taiwan and South Africa have very little collaboration with each other.

INTERVIEWS

We conducted several interviews to contextualise the literature review and bibliometric analysis. We found that South Africa is leading in biotechnology research in certain areas, but there is still a lack of local capital and capabilities to commercialise home-grown biotechnology. As one of the scientists says,

> *Nothing is being commercialised in South Africa that is developed completely from South Africa. It is too expensive and we can't afford it. The trials are also too expensive* (Interview, WS45052).

Similarly, a manager at an innovation hub laments the nature of commercialisation in South Africa. She feels that there is not enough innovation in the country and that if an inventor creates a novel product, it will not receive the proper financial support. She says,

> *The challenge that we find sometimes is that we don't have enough innovations to fund the other one. And you find innovators that have good ideas, good products, but they are not really proper investors. Those are the major challenges but then obviously there are also skills, infrastructure and funding challenges* (Interview, WS450272).

Figure 5.3: Research networks in South Africa and Taiwa

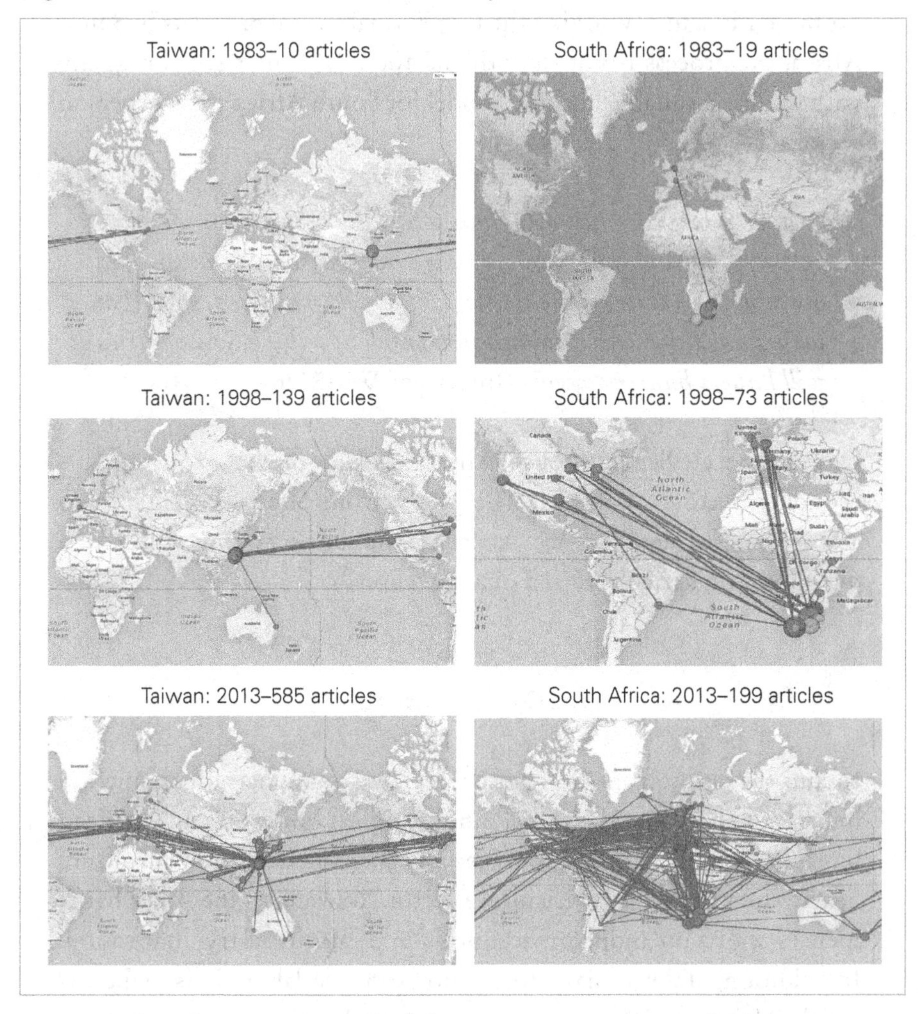

To overcome the funding obstacles, in recent years the South African government has set up funding agencies, such as the TIA, to assist in technology commercialisation and product development, and they have sought funding from a variety of development organisations.

> *Being essentially the main African vaccine manufacture, so you can imagine the stakeholders we deal with, we go to foundations, African Development Bank they provide loans more than a grant, we go to world health organisations* (Interview, WS450264).

Unfortunately, even when the firms receive public funding support to promote PPPs in developing the biopharmaceutical products in South Africa, the process is not smooth and the PPP firm is still struggling with finding a suitable business model for South Africa. As one person says,

> *... it certainly was not a smooth path but nonetheless we have progressed from one into another so we would never find ourselves in a state where we are completely paralysed, so in that way we will have to establish new models; I think South Africa will have a human vaccine* (Interview, WS450264).

Despite the challenges, South Africa has one of the strongest bio-economy sectors in Africa. The R&D from South Africa spreads across the continent and we expect that its bio-economy will have a big impact on agriculture and pharmaceuticals.

CONCLUSION

In sum, the results from the literature review, scientometric mapping and interviews show that the global innovation networks of South African biotechnology R&D are still in the early stages of development and are heavily focused on the agriculture sector. Taiwan, on the other hand, focuses more on biopharmaceuticals and biochemistry innovation. In addition, Taiwan has strong domestic collaborations, whereas South Africa relies heavily on international research collaborations. However, domestic technology commercialisation capacities are still in the process of being developed.

Another issue that arose in South Africa is that most of the intermediaries to encourage innovation, like the TIA, are relatively small and lack sufficient funding to commercialise products. Therefore it is hard for the bio-economy sector to take off.

Despite the challenges we believe that South Africa is on its way to integrating resources and is strengthening the connectedness of the local actors. Therefore, this study suggests that more productive strategies would enhance the research capabilities and diversify the specialities of the local firms, improve the research capacity of local academics, and support these academics in engaging with global innovation networks more broadly. Only by building the local research capacity can the nascent bio-economy in South Africa be successfully developed and transformed.

REFERENCES

Balaguer, A, Luo, Y-L, Tsai M-H, Hung, S-C, Chu, Y-Y, Wu, F-S & Wang, K. 2008. 'The rise and growth of a policy driven economy: Taiwan'. In Edquist, C & Hommen, L, eds. 2008. *Small Country Innovation Systems*, pp 31–70. Cheltenham, UK/Northampton, USA: Edward Elgar Publishing.

Asheim, BT. 2011. *Handbook of Regional Innovation and Growth.* UK: Edward Elgar Publishing.

Bianchi, M, Cavaliere, A, Chiaroni, D & Chiesa, V. 2011. 'Organisational modes for open innovation in the bio-pharmaceutical industry: An exploratory analysis'. *Technovation* 31(1), pp 22–33.

BPIPO. 2011. 'Introduction to biotechnology & pharamceutical industries in Taiwan'. Taipei: Ministry of Economic Affairs.

BOST. 2016. 'Taiwan bio-economy industry development program.' Taipei: Board of Science and Technology.

Breznitz, D. 2007. *Innovation and the State: Political choice and strategies for growth in Israel, Taiwan, and Ireland*, New Haven and London: Yale University Press.

Brinkley, I. 2006. *Defining the Knowledge Economy*. London: The Work Foundation.

Bruhn, M & Gallego, FA. 2012. 'Good, bad, and ugly colonial activities: do they matter for economic development?' *Review of Economics and Statistics*, 94(2), pp 433–461.

Burt, RS. 2001. 'Structural holes versus network closure as social capital, in Nan Lin'. In Cook, KS & Ronald, SB, eds, *Social Capital: Theory and Research'*. New York: Aldine de Gruyter, pp 31–56.

Burt, RS. 2004. 'Structural holes and good ideas'. *American Journal of Sociology,* 110(2), pp 349–399.

Chen, SH, Egbetokun, AA & Chen, DK. 2015. 'Brokering knowledge in networks: Institutional intermediaries in the Taiwanese biopharmaceutical innovation system'. *International Journal of Technology Management*, 69(3–4), pp 189–209.

Chu, W-W. 1997. 'The "East Asian Miracle" and the Theoretical Analysis of

Industrial Policy: A Review', Academia Sinica, Taipei. Available online at: www.idv.sinica.edu.tw/wwchu/SURVEY.pdf (accessed on 26 January 2014).

Central Intelligence Agency. 2016. 'CIA – The World Factbook.' Available at: www.cia.gov/library/publications/the-world-factbook/geos/sn.html (accessed on 26 April 2016).

Coleman, JS. 1988. 'Social capital in the creation of human capital'. *American Journal of Sociology*, 94, S95–S120.

Coleman, JS. 1994. 'Social capital, human capital, and investment in youth'. In Petersen, AC & Mortimer, JT. eds. *Youth Unemployment and Society*. New York: Cambridge University Press, pp 34–50.

Cozzens, S, Rodrigo, C, Harsh, M, Soumonni, D, Wetmore, J & Woodson, T. 2014). 'Nanotechnology and the Two South Africas'. Working Paper, Georgia Institute of Technology.

DCB. 2008. Yearbook of Biotechnology Industry 2008. Taipei MOEA.

DCB. 2011. Yearbook of Biotechnology Industry 2011. Taipei MOEA.

Den Hertog, P & Bilderbeek, R. 2000. 'The new knowledge infrastructure: The role of technology-based knowledge-intensive business services in national innovation systems'. In Boden, M & Miles, I. eds. *Services and the Knowledge-based Economy*. Alberta, Canada: University of Calgary Press, pp 222–246.

Department of Science and Technology. 2001. A National Biotechnology Strategy for South Africa.

Department of Science and Technology South Africa. 2011. Department of Science & Technology Strategic Plan for the Fiscal Years 2011–2016. Pretoria.

Dolfsma, W & Soete, L. eds. 2006. *Understanding the Dynamics of a Knowledge Economy.* UK: Edward Elgar Publishing.

Enriquez-Cabot, J. 1998. 'Genomics and the world's economy'. *Science,* 281, pp 925–926.

Etzkowitz, H & Brisolla, SN. 1999. 'Failure and success: The fate of industrial policy in Latin America and South East Asia'. *Research Policy*, 28(4), pp 337–350.

Fisher, D & Atkinson-Grosjean, J. 2002. 'Brokers on the boundary: Academy-industry liaison in Canadian universities'. *Higher Education,* 44(3), pp 449–467

Freeman, C. 1991. 'Networks of innovators: A synthesis of research issues'. *Research Policy,* 20(5), pp 499–514.

Freeman, LC. 2011. 'The development of social network analysis – with an emphasis on recent events'. In Scott J & Carrington, PJ. eds. *The SAGE Handbook of Social Network Analysis*. London: Sage Publications Ltd, pp 26–39.

Giuliani, E. 2011. 'Networks of innovation'. In Cooke, P, Asheim, B, Boschma,

R. eds. *Handbook of Regional Innovation and Growth*. Cheltenham, UK: Edward Elgar Publishing, pp 155–166.

Granovetter, MS. 1973. 'The strength of weak ties'. *American Journal of Sociology*, 78(6), pp 1360–1380.

Hargadon, A & Sutton, RI. 1997. 'Technology brokering and innovation in a product development firm'. *Administrative Science Quarterly*, 42(4), pp 716–749.

Hargadon, AB. 1998. 'Firms as knowledge brokers'. *California Management Review*, 40(3), pp 209–227.

Howells, J. 2006. 'Intermediation and the role of intermediaries in innovation'. *Research Policy*, 35(5), pp 715–728.

Kaghan, WN. 2001. 'Harnessing a public conglomerate: Professional technology transfer managers and the entrepreneurial university'. In Croissant, J & Restivo, S. eds. *Degrees of Compromise: Industrial Interests and Academic Values*. Albany, New York: SUNY Press, pp 70–100.

Kaplan, D. 2008. 'Science and technology policy in South Africa: Past performance and proposals for the future'. *Science, Technology & Society*, 13(1), pp 95–122.

Kruss, G & Lorentzen, J. 2009. The South African innovation policies: Potential and constraint'. In JE Cassiolato & V Vitorino, eds. *BRICS and Development Alternatives: Innovation Systems and Policies*. New York: Anthem Press.

Keller, W. 2001. International technology diffusion, National Bureau of Economic Research.

Lerner, J. 2004. 'The university and the start-up: Lessons from the past two decades'. *The Journal of Technology Transfer*, 30(1), pp 49–56.

Leydesdorff, L. 2006. 'The knowledge-based economy and the triple helix model'. In Dolfsma, W and Soete, L. eds. *Understanding the Dynamics of a Knowledge Economy*. UK: Edward Elgar Publishing.

Leydesdorff, L & Persson, O. 2010. 'Mapping the geography of science: Distribution patterns and networks of relations among cities and institutes'. *Journal of the American Society for Information Science and Technology*, 61(8), pp 1622–1634.

Leydesdorff, L., Rotolo, D. and Rafols, I. 2012. 'Bibliometric perspectives on medical innovation using the Medical Subject Headings of PubMed'. *Journal of the American Society for Information Science and Technology* 63(11), pp 2239–2253.

Leydesdorff, L & Zawdie, G. 2010. 'The triple helix perspective of innovation systems'. *Technology Analysis & Strategic Management*, 22(7), pp 789–804.

Meyer, M & Kearnes, M. 2013. 'Introduction to special section: Intermediaries between science, policy and the market'. *Science and Public Policy*, 40(4), pp 423–429.

Nelson, R. 1993. 'National innovation systems: A comparative analysis'.

University of Illinois at Urbana-Champaign's Academy for Entrepreneurial Leadership Historical Research Reference in Entrepreneurship.

OECD. 1996. *The Knowledge-Based Economy*. Paris: OECD.

Organisation for Economic Co-operation and Development, & OECD International Futures Programme. 2009. 'The bio-economy to 2030: Designing a policy agenda'. Paris: Organization for Economic Co-operation and Development.

Ozawa, T. 2003. 'Pax Americana-led macro-clustering and flying-geese-style catch-up in East Asia: mechanisms of regionalized endogenous growth'. *Journal of Asian Economics,* 13(6), pp 699–713.

Pavitt, K. 1998. 'The social shaping of the national science base'. *Research Policy,* 27(8), pp 793–805.

Powell, WW & Snellman, K. 2004. 'The knowledge economy'. *Annual Review of Sociology*, 30, pp 199–220.

Powell, WW. 2013. 'Dr. Powell's research interests'. Available at: www.woodypowell.com/research-interests (accessed on 27 December 2013).

Scerri, M. 2009. *The Evolution of the South African System of Innovation Since 1916*. New Castle: Cambridge Scholars Publishing.

Senker, J, Reiss, T, Mangematin, V & Enzing, C. 2007. 'The effects of national policy on biotechnology development: The need for a broad policy approach'. *International Journal of Biotechnology*, 9(1), pp 20–38.

Simoes, AJG & Hidalgo, CA. 2011. 'The economic complexity observatory: An analytical tool for understanding the dynamics of economic development'. Scalable Integration of Analytics and Visualization. Paper from the 2011 AAA1 workshop, San Francisco, USA..

South African Communication and Information System. 2011. 'About South Africa – History'. *South African Yearbook*. Available at: www.info.gov.za/aboutsa/history.htm (accessed on 5 May 2011).

Staffas, L, Gustavsson, M, & McCormick, K. 2013. 'Strategies and policies for the bio-economy and bio-based economy: An analysis of official national approaches'. *Sustainability*, 5, pp 2751–2769.

Stiglitz, J. 1999. 'Public policy for a knowledge economy'. Remarks at the Department for Trade and Industry and Center for Economic Policy Research, 27 January, London.

STPI, ed. 2012. *Science and Technology Yearbook 2012*. National Science Council (NSC), Taipei, Taiwan.

Sun, JC-L. 2005. The National Innovation System of Taiwan's Biotechnology Industry, University of Cambridge.

Tether, BS & Tajar, A. 2008. 'Beyond industry-university links: Sourcing knowledge for innovation from consultants, private research organisations and the public science-base'. *Research Policy,* 37(6–7), pp 1079–1095.

Vallas, SP & Kleinman, DL. 2008. 'Contradiction, convergence and the knowledge economy: The confluence of academic and commercial biotechnology'. *Socio-Economic Review,* 6(2), pp 283–311.

Verma, AS, Agrahari, S, Rastogi, S, Singh, A. 2011. 'Bio-technology in the realm of history'. *Journal of Pharmacy and Bioallied Sciences*, 3(3), pp 321–323.

Wang, C-LJ. 2011. 'Retrospect and prospect of the evolution and innovation of biotech & pharmaceutical industry'. *Science Development*, 45(7), pp 146–149.

White House. 2012. 'National bio-economy blueprint'. Washington DC: The White House.

World Bank. 2011. 'World Bank: Open Data.' Washington D.C. World Bank. Available at: www.data.worldbank.org/ (accessed on 30 August 2014).

World Bank. 2012. 'How we classify countries'. Data. Available at: www.data.worldbank.org/about/country-classifications (accessed on 15 August 2013).

World Bank. 2014a. 'Databank – World Development Indicators – Agriculture'. Available at: wdi.worldbank.org/table/4.2#, (accessed on 30 November 2014).

World Bank. 2014b. 'World Bank: Open Data.' Washington D.C. World Bank. Available at: www.data.worldbank.org/ (accessed on 6 September 2014).

Yusuf, S. 2008. 'Intermediating knowledge exchange between universities and businesses'. *Research Policy*, 37(8), pp 1167–1174.

Zook, MA. 2004. 'The knowledge brokers: Venture capitalists, tacit knowledge and regional development'. *International Journal of Urban and Regional Research,* 28(3), pp 621–641.

SIX

What can South Africa learn from high technology patents in India

An analysis of biotechnology patents through USPTO

SWAPAN KUMAR PATRA & MAMMO MUCHIE

INTRODUCTION

SCIENTIFIC RESEARCH AND TECHNOLOGICAL development is crucial for the economic growth and for the social welfare of a nation. However, major technological innovation is generally emerging from industrially developed countries. In today's globalised world, research and development (R&D) in high technology sectors is carried out in well-equipped sophisticated government laboratories or R&D intensive multinational enterprises (MNEs). So, innovation may be considered a typical 'first-world activity' (Fagerberg et al, 2011). It seems that developed countries are always developing cutting-edge technologies and are at the forefront of technology leadership while developing countries are continuously lagging behind. As such, a few countries are at the forefront of the rapid technological and organisational change that has happened in the recent years, while others at the periphery of the new 'technological paradigm' (Lall, 1998). The technological gap between developed and developing countries has traditionally been regarded as a major reason for underdevelopment and dependency on foreign technology. However, there are exceptions to this hypothesis. During the period between the 1970s and 1980s some of the East Asian 'tigers', for example newly industrialised Singapore and South Korea, were able to reduce the technological gap through careful policy in their respective countries (Kim, 1980). These countries achieved their technological capability by selectively supporting technological development in some key sectors of their economies. Scholars in these areas have suggested that, in the case of new technologies, developing countries might 'leapfrog' into accelerated development, if emphasis is given to the relevant economic actors (Gonsen, 1998).

Against this background, this paper is an attempt to assess the 'technological capability' at a national level in two emerging and developing economics, namely India and South Africa. Both India and South Africa have similar socio-economic and demographic conditions, and both countries have a colonial legacy. They have a similar set of institutions, for example the Council of Scientific and Industrial Research (CSIR) and the Department of Science and Technology (DST). Governments in both countries fund these chains of institutions to look after the overall science and technology (S&T) development in the country. Both countries have a similar structure of universities and tertiary systems of higher education. However, scientific and technological development in India started in the early 1950s, after independence. India is doing substantially well in terms of high technology capability development in the recent years. Perhaps South Africa is not on par with high technology development, particularly information and communication technologies (ICT) and biotechnology areas. The major focus of this chapter is what South Africa can learn from the Indian experience. The following section of this chapter discusses the analytical framework, particularly the literature on technological capability building. Thereafter the chapter deals with the S&T policies of both countries for a better understanding of the technological development trajectories in each. This is followed by a discussion of the objectives and then the detailed methodology.

Subsequently, the chapter details the empirical findings based on the patent analysis from the United States Patents and Trademark Office data (USPTO). The chapter ends with concluding remarks and policy implications.

TECHNOLOGICAL CAPABILITY BUILDING

The 'old' neoclassical growth theory (Solow, 1956) was based on the idea that science and technology have 'public good' characteristics than can be transferred into a global market place. Further, this theory assumed that technology and all its alternatives are freely available everywhere to anyone (Nelson, 1987). Interested people all over the world can use and assimilate the technology without any cost. This theory also further emphasised that the new technologies will always develop in advanced countries because of their well-equipped laboratories and it will steadily diffuse to the developing part of the world. So by the process of diffusion, developing countries will catch up with developed countries. Neoclassical economics, particularly the models proposed in 'trade theory', did not consider technological learning and adaptation as an important issue, particularly in developing countries. Rather than focusing on the need for domestic technological capability building, this theory tended more towards 'north–south' technology transfer (Mytelka, 2003; Morrison et al, 2008). From these schools of thought, the learning process will be more 'imitative' than 'innovative' and involving 'catching up'. Selective interventions by the government can be justified within this neoclassical frame work, if market failure is taken into account (Lall, 1992). In this way global technological development can act as a powerful 'equilibrating force' in the global economy (Fagerberg et al, 2011). However, in reality this did not happen and some countries are always at the technological frontier and many countries are always lagging behind with no sign of catching up (Archibugi & Coco, 2004; Fagerberg et al, 2011).

As neoclassical growth theory did not account for this, a new strand of literature appeared during the 1980s and 1990s (new growth theory). A new stream of literature called the 'technological capability' approach represents a radical alternative to the neoclassical framework (Morrison et al, 2008). The microlevel analysis of technology in developing countries has drawn inspiration from the 'evolutionary theories' developed by Nelson and Winter (1982). In the context of developed or less developed countries, the evolutionary approach is a far more reasonable alternative than the earlier production function approach (Lall, 1992). These new streams of literature argued that effective utilisation of available technology for national development requires some kind of 'capabilities'. The term 'technological capability', was originally coined as a tool to examine the growth of Korean firms by Linsu Kim. However, the term gradually became more widely accepted and much scholarly literature appeared which drew on this approach to understand the performance of different entities at the different levels (Fagerberg & Godinho, 2004; Fagerberg et al, 2011). The concept 'technological capability' is understood from different perspectives, for example at firm, industry, and country level. It involves the ability to adopt and use technologies (Kim, 1997: 4); and the ability to manage technological changes (Bell, 1984). More precisely, technological capability is the ability to generate, use, adapt or change existing technologies. Through this process, firms can move into different levels of technological capability – basic, intermediate and advanced technological capabilities (Lall, 1992; Ariffin & Figueiredo, 2003) depending on their *absorptive capacities* (Cohen & Levinthal, 1990).

The concept 'technological capability' is increasingly being used in different industries and even to analyse the capability of countries. Lall (1992) considered national technological capabilities (NTC) in developing countries to be the accumulation of thousands of individual firm-level capabilities. However, national-level capabilities can be categorised into three types. The categories are: physical investment, human capital and technological. The 'national technological effort' is related to measures taken. For example, on the input side it could be R&D expenditure, or increased numbers of scientific and technical personnel; on the output side it could be the patents filed or granted in national and foreign patent offices, scientific literature published in the peer-reviewed scholarly journals (for example, journals indexed in Web of Science (WoS) of Thomson Reuters or Scopus of Elsevier). Moreover, national technological capability does not only depend on domestic technological efforts but also on foreign direct investments (FDI) and the technologies transferred by the foreign firms engaged in manufacturing or R&D in an offshore location. The 'new growth theory' also further supports the idea that small countries that are at a disadvantage in innovation may come up the value chain through technology transfer by foreign firms in the host countries (Coe & Helpman, 1995; Fagerberg & Srholec 2008; Coe et al, 2009).

It is perhaps difficult to measure national-level technological capability precisely. However, there are proxy measures available to assess the general level of technological capabilities within a country (Manzini, 2015). As previously stated, there are many parameters to measure the input and output side of national technological efforts. On the input side, national level technological capabilities can be measured by the level of domestic R&D expenditure (gross domestic expenditure on R&D (GERD), GERD as a percentage of GDP), the scientific and technical personnel available for R&D, for example (total researchers, total R&D personal and so on) and technical workers. On the output side, the indicators are innovation counts, patents and other indicators of technological success (Lall, 1990; 1992). However, it is important to note here that not all effort is equally effective, and no one measure fully captures routine engineering and incremental innovation. Hence, a combination of different indicators may perhaps yield a better result for a suitable conclusion (for a detailed review see, Archibugi & Coco, 2004; 2005; Fagerberg et al, 2011).

MEASURE OF TECHNOLOGICAL CAPABILITY THROUGH PATENTS ANALYSIS

As per the definition discussed earlier, technological capability may be considered as 'the ability to develop, search, absorb and exploit knowledge commercially'. Innovation capability is one important element of innovation capability (Kim 1997; Fagerberg et al, 2011). There are many databases available for capturing S&T capabilities of countries (WoS or Scopus database) and for assessing technological capabilities (patents granted or filed in the respective countries offices). These patent records are freely available from the respective countries' offices; beside this there are many commercially available databases with patent data.

Patents, as measures of commercially generated technological innovations, are widely accepted in scholarly literature (Archibugi & Coco, 2004; Smith, 2006). A large number of scholars have used patent data as a proxy indicator for measuring innovation (Acs & Audretsch, 1989; Belderbos, 2001; and many more). Patent data contains information that is useful to keep track of the technological capabilities of firms and countries. Moreover, patent data is a useful source of comparison, because patent data is stored for a long period and the databases are continuously updated and upgraded (Griliches, 1990; Callaert et al, 2006).

Patents are the technical documents issued and published by various governments all over the world. A patent gives exclusive rights to its holder for the production, application or utilisation of a novel apparatus or process for a specific period of time. The detailed technological description is available in the main text of the patent document. For example, a patent document provides information about the data of a file or grant, the name of the assignees, the name of the inventor(s), the patent class into which the patent may be categorised and so on. Bibliographic data available with the patent documents is a rich source of information and helps to analyse innovation and the innovation process. Using this bibliographic information, various research on innovation is possible (Callaert et al, 2006; Nagaoka et al, 2011).

However, it is important to mention that the patent as an economic indicator has many pros and cons. These have been broadly discussed in many scholarly articles (for details, see, Griliches, 1990; Patel & Pavitt, 1995). Briefly, all inventions are not patentable, patent filing is very costly and many inventors in developing countries cannot afford the cost of patenting and many patents are not commercialised.

Developed countries have systematic and well-organised data on science and technology activities (for example, patent and publication data are widely available). This is not the case in developing and emerging economies. Two indicators to measure the scientific and technological capabilities of developed countries are under-utilised: statistics from the United States Patent and Trademark Office (USPTO) and from Thomson Reuters Web of Science (WoS). USPTO data is available free of cost while WoS (now owned and maintained by Clavirate Analytics) is subscription-based. Researchers may utilise these databases for varied research purposes (Albuquerque, 2005). Among the various patent information resources, patents granted in USPTO are used in many studies. Since the US is the largest and most technologically developed market in the world, it is generally assumed that important inventions and innovations are legally protected in the US market (Archibugi & Coco, 2004). This study extracted patents of the respective countries (i.e. India and South Africa) using the respective country name as the 'inventor's address'. The patent records are downloaded in Excel sheets and grouped in different categories based on the technology classes, assignee and the yearly growth of patents for further analysis based on the research objectives.

INDIAN AND SOUTH AFRICAN GOVERNMENTS' BIOTECHNOLOGY POLICIES

After independence in 1947, the Indian government passed selective policies over a span of time for a suitable and conducive R&D environment in the country (Kumar, 1990; Buckley et al, 2012). The major policies aimed at opening up the economy and encouraging industrialisation were introduced in the 1990s. The most important initiative in this regard is the 'The Statement of Industrial Policy in 1991'. This policy initiative by the Indian government has changed the industrial landscape of the country. The policy shifted towards the creation of more export-oriented, technology-intensive industries (Statement on Industry Policy, 1991). The Indian government's policies further evolved with the opening up of the Indian economy to the rest of the world. The following section deals with the evolution of Indian government policies for the development of biotechnology in India.

EVOLUTION OF THE INDIAN GOVERNMENT'S POLICIES FOR THE DEVELOPMENT OF BIOTECHNOLOGY IN INDIA

India is one of those developing countries that have realised the huge potential of biotechnology for sustainable development and economic growth (Nagaratnam, 2001). Like other developing and emerging economics, the promotion of the biotechnology industry was promoted through 'public policy' rather than any individual firm's initiatives. During the 1980s and until the mid-1990s the major focus of the government was to build scientific manpower and institutions. However, after economic liberalisation, the industry's landscape changed with the entry of many foreign multinationals (MNEs) (Reid & Ramani, 2012).

India, because of its rich biodiversity and vast knowledge base, is particularly suited to be the country of choice for biotechnology activities. It has perhaps one of the most valuable assets, a huge reservoir of highly skilled manpower available at comparatively lower cost. According to an estimate from the Indian Department of Science and Technology (DST), in 2010, an estimated 4 410 000 people were employed in various R&D units in India including in-house R&D units of public and private sector industries. Of these, about 1 930 000 (43.7%) were employed in areas directly related to R&D activities and about 1 240 000 (28.2%) were engaged in auxiliary activities, for example in administrative and non-technical services. The DST estimate further noted that out of the total of 16 093 PhD holders about 8 302 (51.6%) were from the S&T discipline during 2010–11 (Department of Science and Technology, 2013). India, therefore, has a huge base of skilled English-speaking people as almost all universities, businesses and research institutes use English as a medium of instruction, teaching and research (India Economic News, 2003; Peet, 2005; Patra & Chand, 2005).

The Indian government had been paying special attention to the development of biotechnology since the early 1980s. In 1982, it established the National Biotechnology Board (NBB). The NBB was the top government body responsible for identifying priority areas and formulating long-term strategies for the development of biotechnology. NBB was a high-level committee represented by scientists from different S&T institutions and universities and headed by the science member of the Indian Planning Commission. In 1983, the NBB adopted the 'Long Term Plan in Biotechnology for India' to determine the priorities for biotechnology in India. The priorities were aligned with national objectives, including job creation, poverty reduction, environmental protection, food and energy security, healthcare, economic growth and employment generation. In 1986, NBB was converted to an autonomous body named the Department of Biotechnology (DBT) (Chaturvedi & Srinivas, 2014). The initial investment for the creation of the DBT was about US$210 million to support several research institutions across India (Arora, 2005; Chakraborty & Agoramoorthy, 2010). Now the DBT is the national agency that plays a major role in promoting and developing institutional infrastructure, human resources, biotechnology start-ups, biotechnology incubators and industry–academia interactions (Nagaratnam, 2001).

Further realising the importance of biotechnology during the year 2007–08, the Indian government adopted a 'National Biotechnology Strategy'. The policy document put emphasis on the coordination of various R&D agencies, the application of a biotechnology regulatory framework, the promotion of biotechnology industry, international research partnership and human resource development in biotechnology and so on (Chaturvedi & Srinivas, 2014; Department of Biotechnology, 2015).

The DBT is also the apex body that continuously monitors the quality of human resources programmes. In 2012–13, a number of postgraduate courses on biotechnology were being offered at 70 universities across the country (Department of Biotechnology, 2013). The Department of Science and Technology (DST) and the Department of Biotechnology (DBT) were the two major players contributing nearly 50 per cent of the extramural R&D support in the country (Department of Science & Technology, 2013; Department of Biotechnology, 2015). However, besides the DST and the DBT, the following major organisations in India were, and still are, also responsible for the development and support biotechnology research.

These are, the Department of Scientific and Industrial Research (DSIR), the Council of Scientific and Industrial Research (CSIR), the Indian Council of Agriculture Research (ICAR), the Indian Council of Medical Research (ICMR), and the University Grants Commission (UGC).

In recent years both the provincial and federal governments have established biotechnology parks in different parts of the country (Patra, 2014). The Indian government has also taken initiatives to promote industry–academia relationships. To ensure biosafety, risk assessment and regulation, India has recently introduced a bill to establish a Biotechnology Regulatory Authority of India (Chaturvedi & Srinivas, 2014).

India is making significant progress in different branches of biotechnology. Due to the government's continuous efforts since the 1980s, India is now a major player in the field of biotechnology. The Indian biotechnology industry is experiencing annual increases in revenues (Chaturvedi & Srinivas, 2014). According to the *Biospectrum Able Survey*, the Indian biotech industry earned 25 165 crore in revenue (6.98% growth rate). According to that survey, Indian biotechnology has huge potential to be a global hub for biotechnology R&D and manufacturing (Biospectrum Able Survey, 2014). Many leading Indian firms have commercialised 'off patent' generic versions of many popular and highly profitable drugs originally developed by major global pharmaceutical firms. There are also many firms actively engaged in innovation and, in addition, many start-ups have emerged as contract research organisations (CROs) to provide contract research services to many foreign MNEs. This can be observed from the increasing number of clinical trials conducted by the Indian institutions along with many big foreign MNEs. Hence, the contract CRO industry opens up opportunities for Indian firms or institutions (Reid & Ramani, 2012; Chaturvedi & Srinivas, 2014).

However, a couple of issues need to be addressed for India to better harness the biotechnology sector. Reid and Ramani (2012) argued that India has 'strong scientific and technological capabilities', but it is constrained by the weak 'social capabilities' of its labour force, lack of 'institutional capabilities' in regulation and financing, infrastructural constraints and an absence of national programmes to set concrete targets in terms for biotechnology innovations that will promote a more inclusive development (Reid & Ramani, 2012). The study anticipated a selection of research areas to promote such as generic medicines, vaccines and so on will lead to a global positioning for the country. The study further recommends policies to encourage foreign investment and to promote knowledge sharing. Also, India needs to be a solid world-class infrastructure to utilise the opportunities presented by the CRO industry. Further, patenting activities are the key to attracting venture capital funding and foreign capital investments. Consequently, the study recommended patenting activities should be seriously considered by the Indian players (Reid & Ramani, 2012; Chaturvedi & Srinivas, 2014).

SOUTH AFRICAN SCIENCE AND TECHNOLOGY POLICY

Since the end of the apartheid era in 1994, South Africa has made enormous economic progress (OECD, 2007). In the mid-1990s, the newly elected democratic government recognised the role of S&T in national development (Kaplan, 2004; 2008). New initiatives were started to identify key priority areas and to promote selective technology areas for the development of S&T in South Africa for the benefit of society as a whole rather than just the white minority. The government's ultimate goal was to facilitate employment generation, poverty reduction and overall economic growth. The ultimate plan was to build the nation and to be the leader in overall African development.

The government adopted the 'White Paper on Science and Technology' in 1996. The White Paper mostly focused on the building of a National System of Innovation (NSI). The policy blueprint of the 1996 White Paper led to the formation of the National Research Foundation (NRF) in 1998. At the same time, the National Advisory Council on Innovation (NACI) was also established. The Department of Science and Technology (DST) was established as a stand-alone ministry (separate from Arts and Culture) in 2002 to 'provide leadership, an enabling environment, and resources for science, technology and innovation in support of South Africa's development' (OECD, 2007). The DST acts as a nodal agency for the government research organisations, for example, the Council for Scientific and Industrial Research (CSIR). The agency also looks after funding agencies, for example, the National Research Foundation (NRF). Besides these two sources of competitive funds for R&D, the Innovation Fund (1997) and the Biotechnology Regional Innovation Centres (2001), were started. However, it was observed that the South African innovation system reflects some of the characteristics of its colonial legacy. NIS in South Africa is quite developed in some respects, but it has yet to be inclusive. An integrated and coherent framework for innovation and learning which includes tight links between firms and knowledge generating institutions is still lacking (Muchie, 2003).

The Organization for Economic Cooperation and Development (OECD) conducted a study to evaluate the South African National System of Innovation (NSI) in 2006/2007 (OECD, 2007). The study observed that post-apartheid South Africa has successfully opened up its economy to the outside world. The OECD report also observed that many socio-economic challenges still exist. For example, large numbers of people are still excluded from the formal economy. The poverty and unemployment rate is still increasing. The report further observed that South Africa is now in the midst of two specific economic transitions. The country is moving its economy from less dependence on primary resource production and associated commodity-based industries towards greater globalisation (OECD, 2007).

Following the OECD review of South Africa's innovation systems, in July 2010, the Minister of Science and Technology appointed a Ministerial Review Committee to assess the South African science, technology and innovation landscape. The aim of this review was to explore and recommend a suitable NIS design to meet the needs of the country. The committee set up to examine this issue further investigated the NSI:

With respect to its readiness to meet the needs of the country,

> *the extent to which the country was making optimal use of its existing strengths, and the degree to which the country was well positioned to respond rapidly to a changing global context and to meet the needs of the country in the coming ten to thirty years* (DST, 2012: 15).

The Ministerial Review Committee report observed that there are many gaps in the NSI in the effort to shift from a resource-based economy to a more knowledge-intensive economy. The Committee observed these weakness and came up with policy recommendations (DST, 2012).

SOUTH AFRICAN BIOTECHNOLOGY POLICIES

During apartheid, South Africa was mostly isolated from the global S&T landscape. Throughout that period of isolation, domestic S&T capabilities were developed in strategic sectors related to defence, mining and energy. Biotechnology was not the government's priority and a very small level of biotech research was conducted into first and second generation biotechnology. South Africa's biotechnology industry was in its early stages of development and received little or no government support (Cloete et al, 2006). However, some significant progress was made during that period in the medical and healthcare sectors. For example, scientists from South Africa were the first to develop techniques to perform a human heart transplantation (Christian Barnaard in 1967) and the Computed Axial Tomography (CAT) scan was co-developed by the South African physicist Allan Cormack.

After the end of apartheid, the new democratic South African government encouraged the development of a globally competitive biotechnology industry. The government's interest in biotechnology gained substantial momentum after 1994. In 2001, the South African government published its National Biotech Strategy (NBS) and allocated R400 million (US$52.8 million) to create a realistic and long-term plan for its future biotech industry (Akermann & Kermani, 2006). In 2002, NBS established biotechnology regional innovation centres

(BRICs) to develop and commercialise the biotechnology industry. In 2008 BRICs were replaced by the Technology Innovation Agency (TIA), which was an important component of DST's 10-year plan. The TIA's aims were to develop South Africa's biotechnology industry's ability to transfer local research and development into commercial products and services (Uctu & Essop, 2013). However, the initial plan did not succeed, with little achieved in the commercialisation of biotechnology. According to a survey conducted in 2003, it was observed that about 1 000 biotech-related research projects were being carried out in the country and very few products from these projects had actually been commercialised (Akermann & Kermani, 2006; Gastrow, 2008). In more recent years, the government has taken the initiative to push the science-based Bio-Economy Strategy. The plan was approved by Cabinet in November 2014 and launched by Science and Technology Minister (www.brandsouthafrica.com).

South Africa's biotechnology R&D investment is comparatively small by global standards, but South Africa's initiatives are some of the first initiatives in Africa (Gastrow, 2008). The development of biotechnology in South Africa has mainly been in the areas of agricultural biotechnology. The industrial and pharmaceutical sectors are still in the developing stage. This is not the case in South Africa alone. Developing the policies necessary for the progress in biotechnology has been a great challenge not only in the South African context but for most African countries (Andanda, 2009).

OBJECTIVES

Based on the literature review and the analytical framework of technological capability building, this study now focusses on India's and South Africa's strengths in biotechnology based on the patents granted to Indian and South African inventors in the United States Patent and Trademark office (USPTO) database (Bhattacharya & Patra, 2009; 2010). This study does a detailed analysis of India's capability in high technology, taking into account the patents granted by USPTO in biotechnology. The study further analyses South Africa's position in biotechnology in relation to India's position. In order to assess the position of these two countries, the bibliographic patent data was extracted from the USPTO website and detailed records were extracted from the Thomson Innovation database using South African and Indian inventors' addresses separately. Based on the available information, two separate databases were created for Indian and South African patents. In short, this study analyses the patenting activity of Indian and South African inventors. The analysis focused on growth, the entities involved, and areas of activity, institutional productivity, major actors and the collaboration pattern among the entities.

Limitations of the study

As has already been mentioned in the earlier section, developed countries have well organised and systematic data on science and technology activities. This is not the case of developing and emerging economics. With this limitation in mind, this study considers only the patents granted in the USPTO office. The data from the respective countries, for example the Indian and South African patent offices, technology transfer agreements and other indicators are not considered. Including these indicators may perhaps give an overall picture of the technological capability of both the countries in this high technology industry. Also, this study is based on the four-digit IPC code for biotechnology as suggested by the OECD. A further detailed analysis of IPC codes beyond the four-digit IPC code will show the strength and limitation of biotechnology R&D in India and South Africa.

METHODOLOGY

As discussed, this study uses patents granted to the Indian and South African inventors in USPTO patents as the proxy measure of the technological positions of India and South Africa. This study has examined the available patents for Indian and South African inventors from 1971 to 2014. Based on the available literature, patents of a country can be identified two ways in the USPTO database: 'assignee country' patents and 'inventor country' patents (Bhattacharya, 2004).

For the purpose of this study, patent records were searched where the inventor's address is India or South Africa. In this way, all patents from Indian inventors and South African inventors were searched and retrieved. The details of these patents were further retrieved from the Thomson Innovation database. The data set was categorised into three groups. In case of Indian patents those groups are as follows. In the first group there are 'Indian entity assigned patents', in the second group there are 'unassigned or individually assigned patents' and in the third group there are patents assigned to other than an Indian entity or these can be termed 'foreign assigned patents'. The inventors' addresses were collected by year and collated in an Excel spreadsheet. From this whole data set the patents in biotechnology areas were separated and taken for further analysis. Similar search strategies were applied for South Africa. South African patents are also further categorised into three categories. These categories are 'South African patents' (henceforth called SA patents), 'unassigned or individually assigned patents' and 'foreign assigned patents'.

A high technology industry is defined as one which requires a high proportion of R&D expenditure and employs a high proportion of scientists and engineers (Chakrabarti, 1991). For this study, patents granted in biotechnology industries are considered. Patents in biotechnology industries are identified using the International Patent Classification (IPC) system categorised by OECD (OECD, 2008). IPC codes are part of an internationally recognised patent classification system. The IPC's hierarchical structure denotes sections, classes, subclasses and groups. IPC symbols are assigned according to technical features in patent applications. A patent application can be assigned multiple IPC symbols, as it may relate to multiple technical features.[15]

RESULTS

Search results from the USPTO database for Indian and South African

15 Details of IPC codes for these two industries as suggested by OECD are available at: www.oecd.org/science/inno/37569377.pdf.

inventors show that the earliest records, for both the countries, are available from 1971. For this study the patent records for both the countries are considered from 1971 to 2014. For the abovementioned time period, there are about 19 952 patents granted from Indian inventors and 5 264 patents granted from South African inventors. The next section deals with the overall patenting scenario of India and South Africa. Biotechnology patents were extracted based on OECD-devised IPC codes for biotechnology (for details see Appendix I).

Patent portfolio India and South Africa in USPTO

The overall patent profile of India and South Africa shows that in the USPTO, India has about four times more patents granted than South Africa.The earliest patents can be traced back to the year 1971 for both India and South Africa. In the year 1971 there are about 46 patents available for South African inventors where as there were only five patents available to Indian inventors. However, the greater productivity of South African innovators may be attributed to the fact that they were filing patents to the USPTO. The low number of patents granted to Indian innovators was perhaps due to them filing in the Indian Patent Office, rather than the US. Up to 1998, South African inventors were granted more patents than Indian inventors. However, Indian patenting trends took an upward turn after 1998. This upward growth continues and from 2005 the growth was exponential.

Figure 6.1 shows the comparative yearly growth of Indian and South African patents as available on the USPTO database. It can be observed from this that in recent years India achieved 4 000 patents a year (Figure 6.1). This growth of patenting is due to the two reasons. First, India opened up its economy to foreign investors in the early 1990s. As a result, many foreign firms along with their normal market-related operation have opened up their R&D units in India. Among many other reasons, firms have opened their R&D units in India to take advantage of the 'low-cost high-skilled manpower available at comparatively lower cost' than the developed countries. Second, in 1995 India became a party of the Trade-Related Aspects of Intellectual Property Rights (TRIPS). Because of stricter intellectual property rights, both the domestic and international firms increasesd their R&D

investment. This resulted in an increase in patenting of Indian firms. Domestic Indian firms, particularly Indian pharmaceutical firms, are now becoming more competitive in the global market (Banerjee & Nayak, 2014). In addition, Indian universities and government research institutes (for example, the Council of Scientific and Industrial Research) are quite active in patenting.

The growth rate of South African patents is still static and in the last few years only about 200 patents a year were granted to South African

Figure 6.1 Patent profile of India and South Africa

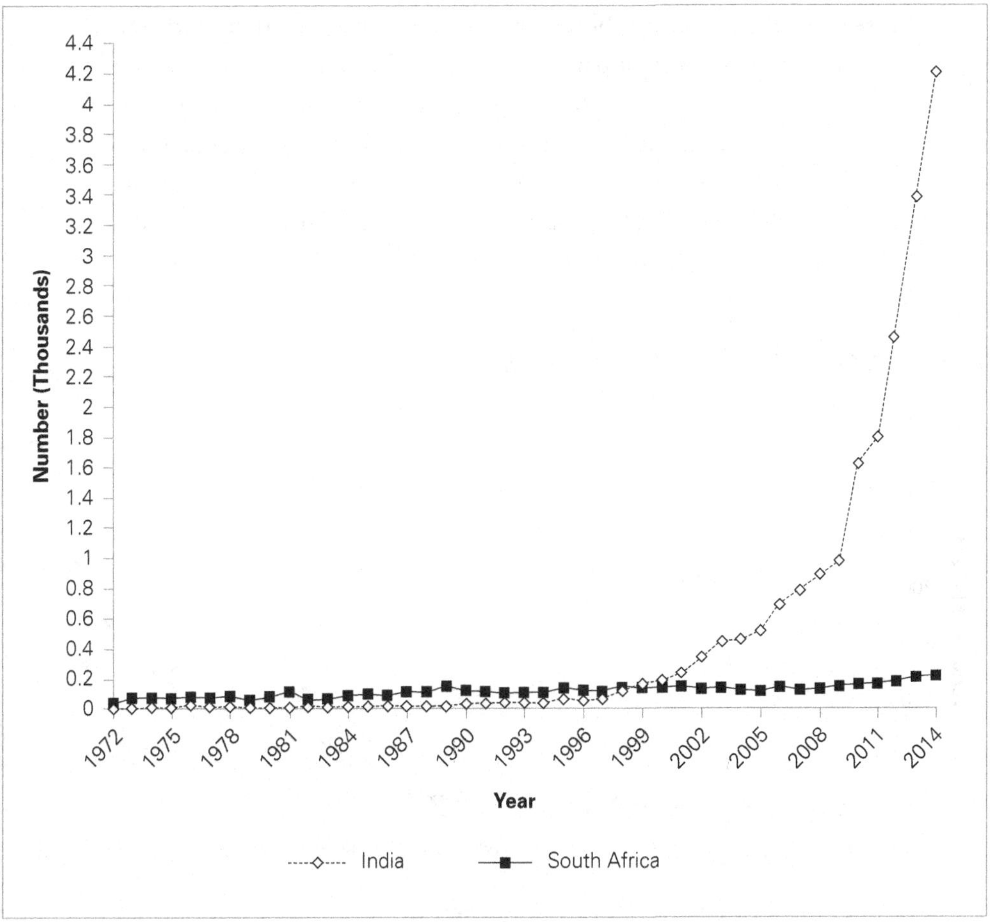

Source: Own compilation from USPTO data

inventors. Although South Africa's Intellectual Property Right (IPR) system is comparatively stronger, it does not perform well. Kaplan (2009) has observed that the South African innovation system is not efficient in terms of innovation outputs compared to the inputs. Although South Africa has a very strong and well-articulated IPR system, Kaplan's study reports that in comparison to other countries, the patenting by MNEs based in South Africa (either domestic or foreign) was very low (Kaplan, 2009).

Biotechnology patents of India and South Africa in USPTO

Indian and South African biotechnology patents are further extracted from the whole sample of patents granted to Indian inventors. Of the 19 954 patents granted to Indian inventors from the year 1971 to 2014, about 2 576 (12% of total patents) are in biotechnology patents based on the codes suggested by the OECD for high technology patents (four-digit IPC codes). The genesis of Indian biotechnology was in the year 1976. In that year about five patents are regarded as biotechnology

Figure 6.2: Growth of biotechnology patents in India

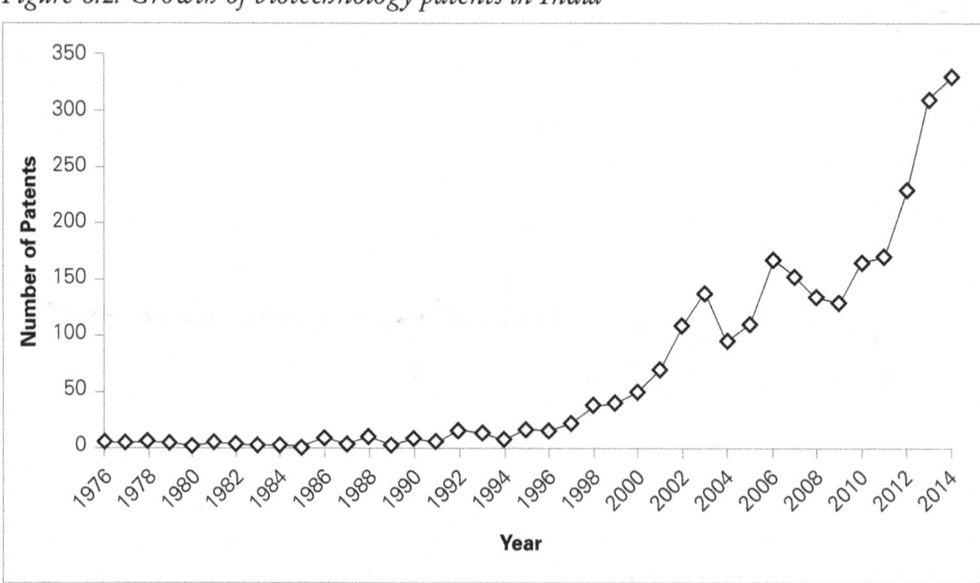

Source: Own compilation from USPTO data

patents. A similar linear type of growth trend is observed until the year 1996. In that year about 15 patent were granted in biotechnology. The number more than doubled in the year 1998 (about 36 patents). After that there was substantial and visible growth except for the year 2004. The reason for the decline of the number of granted patents in 2004 is not very clear. However, from 2013 onwards it can be conluded that more that 300 biotechnology patents were granted to Indian inventors and entities (Figure 6.2). It is apparent from Figure 6.2 that the growth of biotechnology patents started almost after the economic liberations and particularly after 2000. The trend in the last two years indicates a growth of biotechnology patents from India. However, this growth is quite low in terms of other developed countries. According to the OECD Patent Database, the United States contributed 41.5% of all biotechnology PCT (Patent Cooperation Treaty) patent applications during the 2007–09 period. Japan and Germany follow with respective shares of 10.9% and 7.3% (OECD, 2011). According to OECD patent PCT application, India occupies 20th position and South Africa is well behind India in 30th position.

Figure 6.3: Growth of biotechnology patents in South Africa

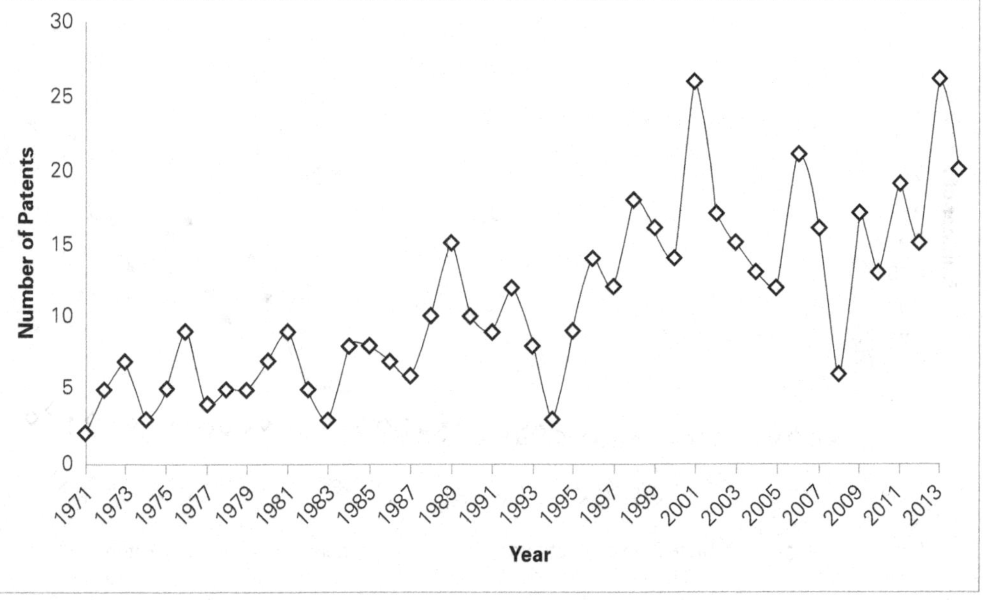

Source: Own compilation from USPTO data

South African activity in biotechnology started quite early. However, the South African patenting in USPTO is not that significant in number. From 1971 to 2014 about 484 patents from South African inventors can be taken as biotechnology patents (Figure 6.3). In the year 1971, two patents were granted to foreign assignees. In the earlier years most of the patents were granted to foreign assignees. After the year 1994, there is a growth of biotechnology patents from South Africa. However, there is no clear and significant trend observed. There was some visible growth for a couple of years and a sharp decline in the consecutive years.

The comparative study of growth of biotechnology patents from these two countries (Figure 6.4) shows that for the study period (1971–2014) there were about 484 patents that were granted to South African inventors whereas 2 576 patents were granted to Indian inventors. Hence, in terms of number, Indian patents are more than five times that of South Africa in the area of biotechnology. Although in the initial years the number of patents from both Indian and South

Figure 6.4: Comparative growth of patents in India and South Africa

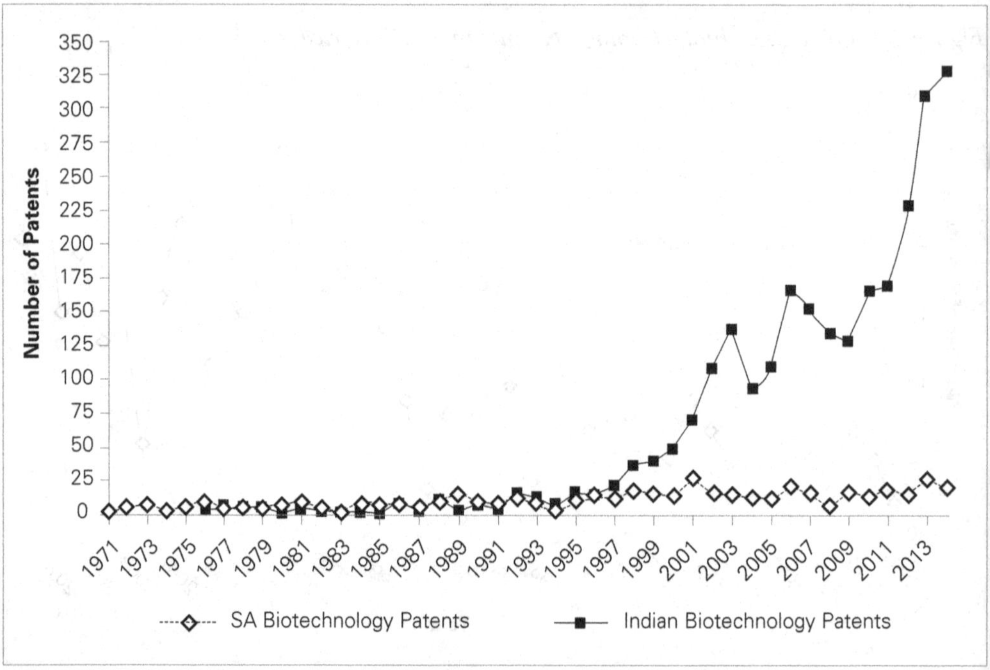

Source: Own compilation from USPTO data

African inventors were almost the same, Indian patenting activities gathered speed after the economic liberalisation of 1991 and gained momentum after the year 2004. From 2013 onwards, Indian entities achieved more than 300 patents per year. In line with patenting trends for the last couple of years, it is expected that Indian patenting will increase in time. In comparison, patenting from South Africa has been less predictable.

Patenting in different classes of technology

The OECD Compendium of Patent Statistics published in 2008 provides patent indicators for international comparisons. Using the four-digit IPC codes, Indian patenting activity can be observed in 11 biotechnology class (Figure 6.5). The IPC codes with more than 200 patents are as follows: A61K, C12N, G01N, C07K and C12P. The rest are other areas (C12Q, A01H, C02F, C12M, C07G and C12S). The maximum number of patents were granted in medical or veterinary science; hygiene (A61K). Among the total 2 577 biotechnology patents, 1 744 (about 68%) can be categorised in this group. The second highest production of patents was observed in C12N (micro-organisms or enzymes). This group has 387 patents (about 15%).

Figure 6.5: Indian patenting activities in different technology classes

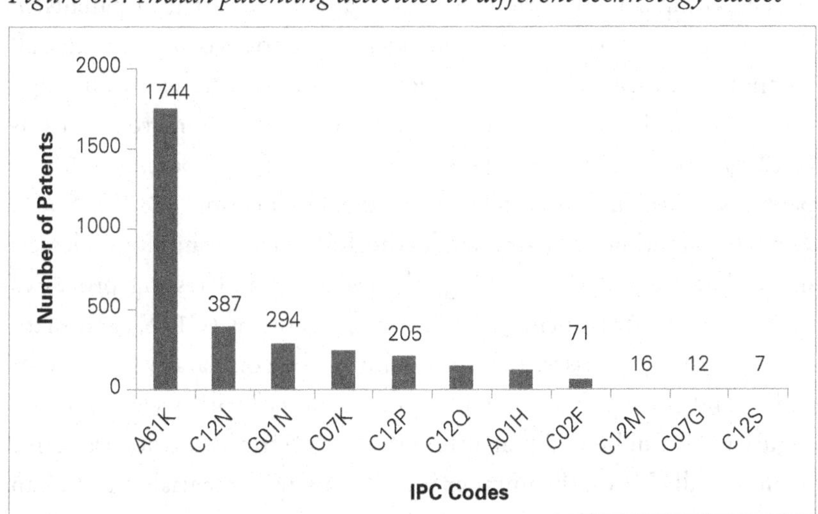

In South Africa, among the total 1 100 high technology patents, 484 patents are granted in biotechnology areas. The patenting activity was mainly observed in 10 technology classes. The IPC codes in terms of higher to lower order are as follows: A61K, G01N, C02F, C12N, A01H, C12P, C07G and C12M. The maximum activity was observed in the technology class A61K (medical or veterinary science; hygiene). There are 174 patents (36 per cent) granted in this group.

Figure 6.6: South African activities in different technology class

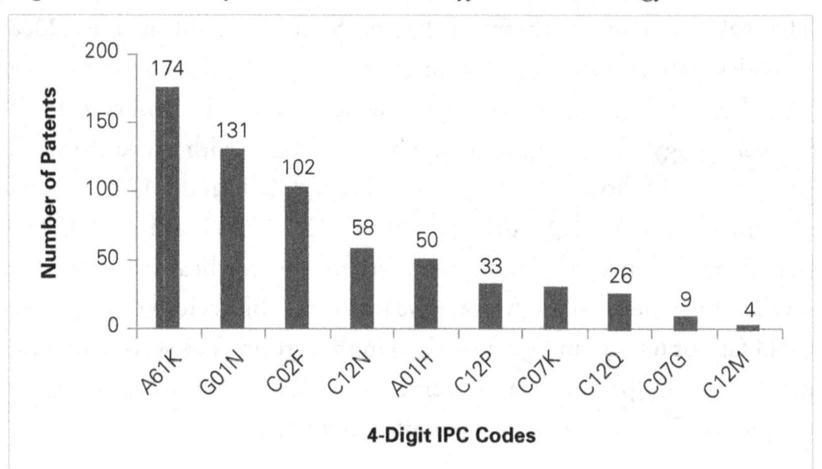

The compartive position of Indian and South African patents in different classes shows that Indian patenting is noteworthy in almost all technology classes. The technological activities in different technology classes show Indian strength in medicine (A61K), microorganisms (C12N), pepdides, biological measuring and testing processes (G01N), peptides (C07K), and fermentation and engineering (C12P). South African patenting activities are in the following technology classes: medical related (A61K), biological measuring and testing processes (G01N), water treatment (C02F), microorganisms (C12N) and plant tissue culture processes. Figure 6.7 shows the comparative picture of Indian and South African patents in different technology classes. From Figure 6.7 it can be said that although South African entities are active in almost all classes, the number is quite low in comparision to Indian patenting activity.

Figure 6.7: Indian and South African patents in different technology classes

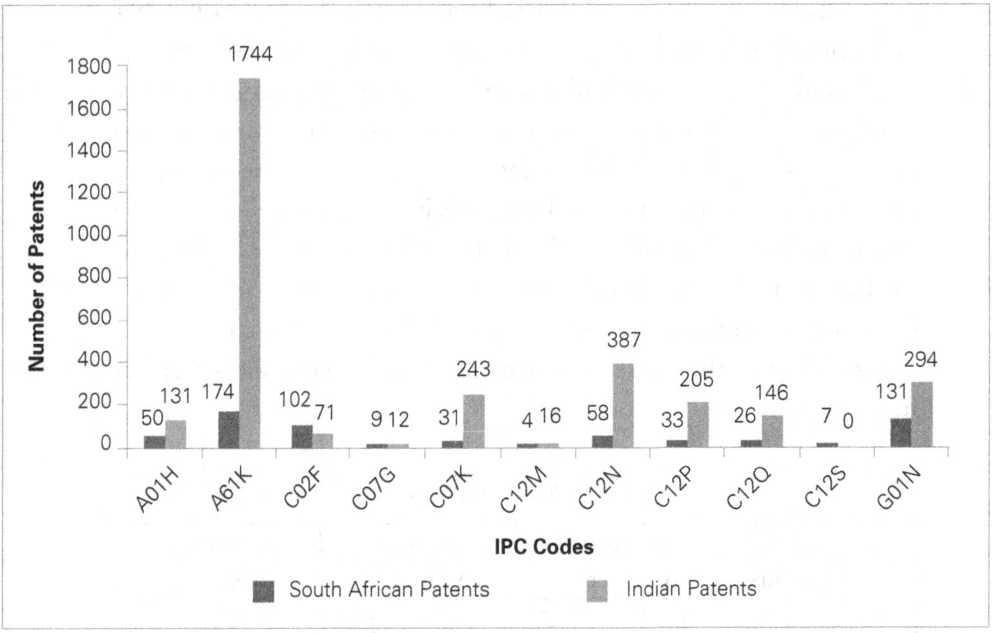

Source: Own compilation from USPTO data

Assignee-wise distribution of patents

In a patent document, an assignee is the owner of the patent. The inventor, generally the employee of a firm, signs over the rights of the invention to the organisation that employs them. Table 6.1 shows the assignee-wise distribution of Indian biotechnology patents. As already discussed in the methodology section, the patents are categorised into three distinct groups. In the first group, there are patents granted to the entities which are solely Indian entities. For example, an Indian entity may be university, firm or institute. The second group consists of patents granted to foreign entities, which include foreign firms, institutes not based in India or foreign universities. The third group consists of individual or unassigined patents because these groups of patents are not granted to any firm or institute. Rather, these patents are granted to individuals. It can be observed in Table 6.1 that Indian entities, firms, universities and government research institutes are quite active in patenting. Table 6.1 shows the total number of patents, categorised under the three distinct groups and the collaborative patents in each group. It can be observed from the table that Indian entities (firms, universities and research institutes) have done quite a good number of patenting in biotechnology. Although the collaboration in patents is generally limited, the Indian entity-assigned patents have many examples of collaborations. The number of collaborative patents has also increased over time. The number of collaborative patents reached an all-time high in recent years. Therefore, it can be concluded from the data in Table 6.1 that collaboration in biotechnology is increasing. For the Indian entities, collaboration happens between firms, and between firms and academia. In the case of foreign MNEs, collaboration mainly occurs between the firm's subsidiary located in India and its global headquarters.

Table 6.1: Assignee-wise distribution of Indian biotechnology patents

	Total Patents	Indian Assignee		Foreign Assignee		Individual Assignees	
		Number of patents	Collaborative patents	Number of patents	Collaborative patents	Number of patents	Collaborative patents
1976	5	-		4		1	1
1977	4	1		2		1	
1978	5	0		5		0	
1979	4	1		3		0	
1980	1	1					
1981	4			4			
1982	3	1		2			
1983	2			1		1	
1984	2			2			
1986	9	1		7		1	
1987	3			1			
1988	10	2		5		3	
1989	3			2		1	
1990	7			5		2	

	Total Patents	Indian Assignee		Foreign Assignee		Individual Assignees	
		Number of patents	Collaborative patents	Number of patents	Collaborative patents	Number of patents	Collaborative patents
1991	5			4		1	
1992	15	1		13		1	
1993	13	1		12			
1994	7	1		6			
1995	16	1		11		4	4
1996	15	1		10		4	1
1997	21	7		9		5	4
1998	36	13		17		6	5
1999	40	20	1	16	3	4	2
2000	49	24	6	17	1	8	6
2001	70	55	6	14		1	1
2002	109	85	6	20	1	4	3
2003	136	108	6	25	3	3	1
2004	94	66	5	20	1	8	6
2005	110	83	4	26		1	1
2006	165	114	8	46	2	5	3
2007	152	92	9	50		10	8
2008	134	85	9	44	1	5	2
2009	129	91	11	33	1	5	2
2010	164	103	7	55	4	6	5
2011	170	112	18	53	2	6	4
2012	228	136	14	82	6	10	5
2013	309	169	13	129	1	11	5
2014	328	190	34	122	34	16	8

Source: Own compilation from USPTO data

Table 6.2 lists the top Indian assignees (with more than 20 patents). The Council of Scientific and Industrial Research (CSIR) is the top assignee with 593 biotechnology patents. CSIR India is the largest government research organisation represented by a chain of about 38 national laboratories. CSIR laboratories employ about 4 600 scientists and about 8 000 scientific and technical personnel. Besides CSIR, other research organisations like the National Institute of Immunology and the Department of Biotechnology are also among the top patentees. Indian pharmaceutical firms are also quite active in patenting. Among the Indian firms, Ranbaxy Laboratories Inc. tops with 83 patents followed by Wockhardt Limited (43) and Dr. Reddy's Laboratories Limited (42).

Table 6.2: Top Indian assignee of biotechnology patents from India

Rank	Name of the Entity	Type	Number of Patents
1	*Council for Scientific and Industrial Research*	Government research institute	593
2	*Ranbaxy Laboratories Inc.*	Firm	83
3	*Wockhardt Limited*	Firm	43
4	*Dr. Reddy's Laboratories Limited*	Firm	42
5	*Cadila Healthcare Limited*	Firm	39
6	*Piramal Enterprises Limited*	Firm	34
7	*Department of Biotechnology*	Government institute	33
8	*Cipla Limited*	Firm	30
9	*Reliance Life Science Pvt. Ltd.*	Firm	29
10	*Lupin Laboratories Ltd.*	Firm	27
11	*Dabur Pharma Limited*	Firm	26
12	*Hetero Drugs Limited*	Firm	25
13	*Indian Institute of Technology*	University	24
14	*Indian Institute of Science*	University	23
15	*Panacea Biotec Limited*	Firm	23
16	*Biocon India Limited*	Firm	22

Rank	Name of the Entity	Type	Number of Patents
17	*Sun Pharamaceutical Industries Ltd.*	Firm	21
18	*Torrent Pharmaceutical Ltd.*	Firm	21
19	*National Institute of Immunology*	Research institute	20

Source: Own compilation from USPTO data

Foreign assignees and their patents are shown in Table 6.3. Indian firms with subsidiaries located in other developed countries are also quite active in patenting. Dr. Reddy US Therapeutics, Inc., part of Dr. Reddy's Discovery Research, is the top patentee. As the patent assignee address shows the address other than Indian address, this firm is counted as a foreign firm. This major Indian pharmaceutical company has established its presence in the US market and is a fully owned subsiadiary of Dr. Reddy's Lab, located in Atlanta, USA. General Electric is the second most patentee followed by Hoechst AG (in 1999 it merged with French pharmaceutical company Rhône-Poulenc to create the French-German pharmaceutical firm Aventis. Aventis was acquired by Sanofi in 2004). The increasing number of patents by firms in India is due to the recent surge of foreign firms and their R&D activities. India has witnessed a major surge in foreign direct investment (FDI) from US$129 million in 1991–92 to US$36 billion in 2011. This was due to a steady opening up of the economy. After liberalising the economy, the Indian government has nurtured the investment climate by passing a number of FDI-favourable policies. These include hassle-free investment, tax exemptions, suitable infrastructure and other incentives. The Indian government also has been instrumental in helping MNEs to set up R&D centres in India (Krishna, Patra & Bhattacharya, 2012; Patra & Krishna, 2015).

The FDI in high technology sectors shows that India is gradually becoming a favourable destination for many high technology firms, including biotechnology, information technology (IT) and pharmaceuticals. It is emerging as a major centre for cutting-edge R&D

projects for global MNEs. An EIU (2007) survey of 300 executives worldwide found that India was the most popular off-shore R&D destination. The main reason is the huge reservoir of high-quality low-cost Indian manpower (EIU, 2007). Many large MNEs are taking advantage of these huge reservoirs of Indian human resources to enhance their R&D skill locally as well as globally (Satyanand, 2007).

Table 6.3: Top foreign assignees of biotechnology with Indian inventors

Rank	Name of the entity	Country of assignee	Number of patents
1	*Dr. Reddy's Laboratories Inc. & Reddy US Therapeutics Inc.*	USA	59
2	*General Electric Company*	USA	50
3	*Hoechst Aktiengesellschaft*	Germany	48
4	*The United States of America as represented by the Department of Health and Human Services*	USA	31
5	*Glenmark Pharmaceuticals S.A.*	USA	24
6	*Bristol Myers Squibb Pharma Company*	USA	21
7	*CV Therapeutics Inc.*	USA	20
8	*Novartis AG*	Switzerland	16
9	*Astra Aktiebolag*	USA	12
10	*Medivation Technologies Inc.*	USA	12
11	*ABB Research Ltd.*	Switzerland	11
12	*Generics (UK) Limited*	UK	11
13	*International Business Machines Corporation*	USA	11
14	*Monsanto Company*	USA	11
15	*Natreon Inc.*	USA	11
16	*Boston University*	USA	10
17	*Genzyme Corporation*	USA	10

Source: Own compilation from USPTO data

Since the foreign semiconductor firm Texas Instruments established its R&D centre in India in 1985, MNEs' R&D investment has gained momentum. According to an estimate by research firm Zinnov, India now has about 1 031 foreign R&D centres. A recent report jointly published by Battelle India & Federation of Indian Chambers of Commerce and Industry (FICCI) has confirmed this phenomenon (John, 2013). Although the initial focus of many firms was the customisation of their products for the Indian market, many firms in India have now reached maturity and are training skilled talent capable of bringing out products for their global market. After the initial years of establishment in India, many international firms are carrying out a significant portion of their product development in India. More and more foreign firms are allocating their strategically important works to their development centres. In recent years, many international companies' Indian R&D centres are driving the overall growth of parent companies. This is borne out by the fact that their R&D centres in India are the largest outside of their home country. For example, Microsoft, Oracle, IBM, SAP and HP have all claimed that their R&D centre in India is the largest outside of their respective home locations. Many of them have future plans to further expand their existing facilities and infrastructure and to carry forward a significant proportion of their R&D work there. Microsoft India Development Center (MSIDC), started in Hyderabad in 1998, is now the largest R&D centre outside its headquarters in Redmond.[16] The Cisco Global Development Center in Bangalore started its operation in 1995 and now is the largest outside the US.[17] Yahoo! set up a full-fledged R&D centre in Bangalore in 2002. Among the biotechnology and healthcare-related activities, Philips Innovation Campus (PIC) at Bangalore is the largest R&D centre for Philips outside Holland.[18] GE's John F Welch Technology Centre in India is heavily backed by Indian talent. It has about 8 000 scientists and engineers (6 000 engineers in Bangalore and 2 000 in Hyderabad). The technology centre has about 1 000 patents

16 www.microsoft.com/en-in/msidc/
17 www.cisco.com/web/IN/about/company_overview.html
18 www.bangalore.philips.com/html/AboutPIC.html

based on work done in Indian R&D centres in different patent offices all over the world.[19] The centre is working on a wide spectrum of high technology including high-end power generation, aviation equipment and healthcare (Patra & Krishna, 2015).

In contrast, biotechnology patenting in South Africa is dominated by domestic firms or institutes (Table 6.4). Among 484 biotechnology patents, 226 are assigned to South African firms, institutes or other entities. There are about 165 patents granted to foreign assiginees and about 100 (about 21%) are assigned to individual entities. The collaboration pattern shows that of the South African assigned patents about 25 patents (about 5%) are joint patents. Among the 165 foreign patentees, 16 patents (about 3%) are joint patents. The assignee-wise distribution does not exhibit any clear and significant trend. Collaboration among the patentees is increasing but the trend is not significant.

Among the biotechnology assignees from South Africa, the University of Cape Town is the most productive institute with 21 patents, followed by the Council for Scientific and Industrial Research, South Africa. The University of Cape Town is also a very important actor in terms of collaboration. This university has many collaborative patents with different entities.

19 www.ge.com/in/oil-and-gas/JFWTC

Table 6.4: Assignee-wise distribution of South African biotechnology patents

		SA Assignee		Foreign Assignee		Individual Assignee	
Year	Number of patents	Number of patents	Number of collaborative patents	Number of patents	Number of collaborative patents	Number of patents	Number of collaborative patents
1971	2	1		1		0	
1972	5	1		2		2	
1973	7	1		1		5	1
1974	3	2				1	
1975	5	3				2	
1976	9	4					
1977	4	1		2		1	
1978	5	1		2		2	
1979	5	1		2		2	
1980	7	2				5	
1981	9	1		2		6	
1982	5	2		1		2	
1983	3	0		2		1	
1984	8	4	1	4	1	1	
1985	8	4		3		1	
1986	7	4		3		0	
1987	6	2		3		1	

		SA Assignee		Foreign Assignee		Individual Assignee	
Year	Number of patents	Number of patents	Number of collaborative patents	Number of patents	Number of collaborative patents	Number of patents	Number of collaborative patents
1988	10	6		3		1	1
1989	15	5		5		5	2
1990	10	4		3		3	1
1991	9	5		2		2	
1992	12	2		5		5	1
1993	8	6		1		1	
1994	3	2		0		1	
1995	9	3	1	2	1	4	2
1996	14	3	1	8		3	2
1997	12	2		7		3	
1998	18	11		5		2	1
1999	16	5		8		3	
2000	14	7	2	5		2	2
2001	26	14	1	8		4	3
2002	17	6		8		3	
2003	15	5	1	8		2	
2004	13	2	2	8	1	4	1

		SA Assignee		Foreign Assignee		Individual Assignee	
Year	Number of patents	Number of patents	Number of collaborative patents	Number of patents	Number of collaborative patents	Number of patents	Number of collaborative patents
2005	12	2		4		6	
2006	21	17		1		3	2
2007	16	7		4		5	2
2008	6	2		4		0	
2009	17	9	2	9	1	0	
2010	13	11	2	2		0	
2011	19	12	2	6	2	3	
2012	15	7	2	5	1	3	1
2013	26	20	6	10	5	0	
2014	20	17	2	6	4	0	

Source: Own compilation from USPTO data

Table 6.5: Top biotechnology assignees from South Africa

Sl. no		Country of Origin	Number of patents
1	*University of Cape Town*	South Africa	21
2	*CSIR*	South Africa	20
3	*South Africa Inventions Development Corporation*	South Africa	14
4	*Water Research Commission*	South Africa	11
5	*Schering Aktiengesellschaft*	Germany	8
6	*Roecar Holdings*	Netherlands	7
7	*Sasol Industries (Proprietary) Limited*	South Africa	7
8	*South African Medical Research Council*	South Africa	7
9	*Adcock Ingram Limited*	South Africa	6
10	*AE & CI Limited*	South Africa	6
11	*Agricultural Research Council*	South Africa	6
12	*University of Pretoria*	South Africa	6
13	*Farmarc Nederland B.V.*	Netherlands	5
14	*National Energy Council*	Netherlands	5
15	*Warburton Technology Limited*	Ireland	5

Source: Own compilation from USPTO data

CONCLUDING REMARKS

This chapter examined the technological capabilities of India and South Africa through the biotechnology patents granted in USPTO. The study shows that in terms of the number of patents, India is far ahead of South Africa. Based on the OECD-devised IPC codes of high technology patents, this study found a significant increase in high technology patents in India in recent years. The increase in Indian patents is generally observed after the year 1999–2000. This is because of the opening up of the Indian economy to foreign firms. As a result of the open economy policy, many foreign firms have opened up their

R&D centres in India in the last couple of decades. These firms are now benefitting from not only the huge Indian market of increasing middle-class population but also the highly skilled human resources available at comparatively lower cost than in developed countries. Therefore, a policy recommendation for South Africa is perhaps to provide incentives for foreign firms to conduct their R&D operations in the country. Moreover, developing skilled personnel, who could act as an incentive for foreign firms to start an R&D operation in the country should be encouraged.

In the Indian patents portfolio, it was observed that, in the biotechnology sector, among 2 576 patents, 1 565 patents are assigned to Indian entities whereas 876 patents are assigned to foreign entities. This shows that in the biotechnology industry Indian firms and institutions are more active. Indian biotechnology firms can lean on the strength in terms of both domestic operations and quality-driven government research laboratories.

However, the technological capabilities of the two countries studied are based on a single indicator (namely, the USPTO data). The country comparison could have benefitted from the inclusion of other indicators. These include patent records from the respective countries' patent offices, R&D expenditure, R&D skills base and so on. Even patent citation analysis of the patents can be done to study the technological strength of these two nations.

APPENDIX I: OECD-RECOMMENDED IPC CODES RELATED TO BIOTECHNOLOGY

A01H1/00	Processes for modifying genotypes
A01H4/00	Plant reproduction by tissue culture techniques
A61K38/00	Medicinal preparations containing peptides
A61K39/00	Medicinal preparations containing antigens or antibodies
A61K48/00	Medicinal preparations containing genetic material which is inserted into cells of the living body to treat genetic diseases; gene therapy

C02F3/34	Biological treatment of water, waste water, or sewage: characterised by the micro-organisms used
C07G 11/00	Compounds of unknown constitution: antibiotics
C07G 13/00	Compounds of unknown constitution: vitamins
C07G 15/00	Compounds of unknown constitution: hormones
C07K 4/00	Peptides having up to 20 amino acids in an undefined or only partially defined sequence; derivatives thereof
C07K 14/00	Peptides having more than 20 amino acids; Gastrins; Somatostatins; Melanotropins; derivatives thereof
C07K 16/00	Immunoglobulins, e.g. monoclonal or polyclonal antibodies
C07K 17/00	Carrier-bound or immobilised peptides; preparation thereof
C07K 19/00	Hybrid peptides
C12M	Apparatus for enzymology or microbiology
C12N	Micro-organisms or enzymes; compositions thereof
C12P	Fermentation or enzyme-using processes to synthesise a desired chemical compound or composition or to separate optical isomers from a racemic mixture
C12Q	Measuring or testing processes involving enzymes or micro-organisms; compositions or test papers therefore; processes of preparing such compositions; condition-responsive control in microbiological or enzymological processes
C12S	Processes using enzymes or micro-organisms to liberate, separate or purify a pre-existing compound or composition processes using enzymes or micro-organisms to treat textiles or to clean solid surfaces of materials
G01N27/327	Investigating or analysing materials by the use of electric, electro-chemical or magnetic means: biochemical electrodes
G01N33/53*	Investigating or analysing materials by specific methods not covered by the preceding groups: immunoassay; bio-specific binding assay; materials therefore

G01N33/54*	Investigating or analysing materials by specific methods not covered by the preceding groups: double or second antibody: with steric inhibition or signal modification: with an insoluble carrier for immobilising immunochemicals: the carrier being organic: synthetic resin: as water suspendable particles: with antigen or antibody attached to the carrier via a bridging agent: Carbohydrates: with antigen or antibody entrapped within the carrier
G01N33/55*	Investigating or analysing materials by specific methods not covered by the preceding groups: the carrier being inorganic: Glass or silica: Metal or metal coated: the carrier being a biological cell or cell fragment: Red blood cell: Fixed or stabilised red blood cell: using kinetic measurement: using diffusion or migration of antigen or antibody: through a gel
G01N33/57*	Investigating or analysing materials by specific methods not covered by the preceding groups: for venereal disease: for enzymes or isoenzymes: for cancer: for hepatitis: involving monoclonal antibodies: involving limulus lysate
G01N33/68*	Investigating or analysing materials by specific methods not covered by the preceding groups: involving proteins, peptides or amino acids
G01N33/74*	Investigating or analysing materials by specific methods not covered by groups G01N 1/00-G01N 31/00 involving hormones
G01N33/76*	Human chorionic gonadotropin
G01N33/78*	Thyroid gland hormones
G01N33/88*	Involving prostaglandins
G01N33/92*	Involving lipids, e.g. cholesterol

Source: A Framework for Biotechnology Statistics, *available at www.oecd.org/sti/biotech/aframeworkforbiotechnologystatistics.htm*

REFERENCES

Acs, ZJ & Audretsch, DB. 1989. 'Patents as a measure of innovative activity'. *Kyklos,* 42(2), pp 171–180.

Akermann, B & Kermani, F. 2006. 'The promise of South African biotech'.

Drug Discovery Today, 11(21/22), pp 962–965.

Albuquerque, EdMe. 2004. 'Science and technology systems in less developed countries: Identifying a threshold level and focusing in the cases of India and Brazil'. In HF Moed, W Glänzel & U Schmoch, eds. *Handbook of Quantitative Science and Technology Research: The Use of Publication and Patent Statistics in Studies of S&T Systems.* Dordrecht: Kluwer Academic Publishers.

Andanda, P. 2009. 'Status of biotechnology policies in South Africa'. *Asian Biotechnology and Development Review*, 11(3), pp 35–47.

Archibugi, D & Coco, A. 2004. 'A new indicator of technological capabilities for developed and developing countries (ArCo)'. *World Development*, 32(4), pp 629–654.

Archibugi, D & Coco, A. 2005. 'Measuring technological capabilities at the country level: A survey and a menu for choice'. *Research Policy*, 34(2), pp 175–194.

Ariffin, N & Figueiredo, P. 2004. 'Internationalisation of innovative capabilities: Counter-evidence from the electronics industry in Malaysia and Brazil'. *Oxford Development Studies*, 32(4), pp 559–583.

Arora, P. 2005. 'Healthcare biotechnology firms in India: Evolution, structure and growth'. *Current Science*, 89(3), pp 458–464.

Banerjee, T & Nayak, A. 2014. 'Effects of trade related intellectual property rights on the research and development expenditure of Indian pharmaceutical industry'. *Journal of Pharmaceutical Health Services Research*, 5(2), pp 89–94.

Belderbos, R. 2001. 'Overseas innovation by Japanese firms: An analysis of patent and subsidiary data'. *Research Policy*, 20(2), pp 313–332.

Bell, M. 1984. 'Learning and the accumulation of industrial technological capacity in developing countries'. In M Fransman & K King, eds. *Technological Capability in the Third World.* London: Macmillan Press Ltd.

Bhattacharya, S. 2004. 'Mapping inventive activity and technological change through patent analysis: A case study of India and China'. *Scientometrics*, 61(3), pp 361–381.

Bhattacharya, S & Patra, SK. 2009. 'Patent as an indicator of technological capability: Case study based on Indian patenting activity in the biotechnology sector'. [FP]. Paper presented at the 12th International Conference on Scientometrics and Informetrics, Rio de Janeiro, Brazil.

Bhattacharya, S & Patra, SK. 2010. 'Assessing competency of a developing country in high technology: A case study based on Indian patenting activity in the biotechnology sector'. *Collnet Journal of Scientometrics and Information Management*, 4(2), pp 21–34.

Biospectrum Able Survey. 2014. New Delhi: Biospectrum.

Buckley, PJ, Cross, AR & Horn, SA. 2012. 'Japanese foreign direct investment in India: An institutional theory approach'. *Business History,* 54(5), pp 657–688.

Brand South Africa website: South Africa moves to grow bio-economy. 15 January 2014. Available at: www.southafrica.info/business/trends/newbusiness/biotech-150114.htm#.VnqSIE3UiAg#ixzz3v98DMTi4 (accessed on May 2016).

Callaert, J, Looy, BV, Verbeek, A. Debackere, K & Thijs, B. 2006. 'Traces of prior art: An analysis of non-patent references found in patent documents'. *Scientometrics,* 69(1), pp 3–20.

Chakrabarti, AK. 1991. 'Competition in high technology: Analysis of patents of U.S., Japan, U.K., France, West Germany, and Canada'. *IEEE Transactions on Engineering Management,* 38(1), pp 78–84.

Chakraborty, C & Agoramoorthy, G. 2010. 'A special report on India's biotech scenario: Advancement in biopharmaceutical and health care sectors'. *Biotechnology Advances,* 28(1), pp 1–6.

Chaturvedi, S & Srinivas, KR. 2014. 'India survey on biotechnology capacity in Asia-Pacific opportunities for national initiatives and regional cooperation: A report prepared for UNESCO Office, Jakarta', pp 75–83. New Delhi: Research and Information Systems for Developing Countries.

Cloete, TE, Nel, LH, & Theron, J. 2006. 'Biotechnology in South Africa'. *Trends in Biotechnology,* 24(12), pp 557–562.

Coe, DT & Helpman, E. 1995. 'International R&D spillovers'. *European Economic Review,* 39(5), pp 859–887.

Coe, DT, Helpman, E & Hoffmaister, AW. 2009. 'International R&D spillovers and institutions'. *European Economic Review,* 53(7), pp 723–741.

Cohen, W & Levinthal, D. 1990. 'Absorptive capacity: A new perspective on learning and innovation'. *Administrative Science Quarterly*, 35, pp 128–152.

Department of Biotechnology, 2013. *Annual Report of the Department of Biotechnology, Government of India, 2012–13.* New Delhi: Department of Biotechnology, Government of India.

Department of Biotechnology, 2015. *National Biotechnology Development Strategy.* New Delhi: Department of Biotechnology, Government of India. Available at: www.dbtindia.nic.in/national-biotechnology-development-strategy-2015-2020-announced-2/ (accessed on April 2016).

Department of Science & Technology, 2013. *Research and Development Statistics at a Glance (2011–2012).* New Delhi: Department of Science and Technology, Ministry of Science and Technology, Government of India.

EIU. 2007. 'Sharing the idea the emergence of global innovation networks: A report from the Economist Intelligence Unit'. London: Economist Intelligence Unit.

Fagerberg, J & Godinho, MM. 2004. 'Innovation and catching-up'. In DMJ Fagerberg & R Nelson, eds. *The Oxford Handbook of Innovation,* pp.

514–544. Oxford: Oxford University Press.

Fagerberg, J & Srholec, M. 2008. 'National innovation systems, capabilities and economic development'. *Research Policy*, 37(9), pp 1417–1435.

Fagerberg, J, Srholec, M, & Verspagen, B. 2011. 'Innovation and economic development'. In BH Hall & N Rosenberg, eds. *Handbook of the Economics of Innovation*, pp. 833-872. Amsterdam, The Netherlands: Elsevier.

DST. 2012. 'Final report of the ministerial review committee on the science, technology and innovation landscape in South Africa'. Department of Science and Technology, Republic of South Africa.

Gastrow, M. 2008.' Great expectations: The state of biotechnology research and development in South Africa'. *African Journal of Biotechnology*, 7(4), pp 342–348.

Gillbert, G. 2005. 'Can the biotech industry takes off in SA?' *Science in Africa*. Available at: www.scienceinafrica.co.za/2005/march/biotechindustry.htm (accessed on May 2016).

Gonsen, R. 1998. *Technological Capabilities in Developing Countries: Industrial Biotechnology in Mexico*. Basingstoke: Macmillan Press.

Griliches, Z. 1990. 'Patent statistics as economic indicators: A survey'. *Journal of Economic Literature*, 28, pp 1661–1707.

India Economic News. 2003. 'Biotechnology in India: A promising future', *India Economic News*, 13, pp 1–6.

John, S. 2013. 'MNC 25 global companies set up R&D centres in India in last 18 months'. Available at: www.timesofindia.indiatimes.com/business/india-business/25-global-companies-set-up-RD-centres-in-India-in-last-18-months/articleshow/22846741.cms (accessed on May 2016).

Kaplan, D. 2004. 'South Africa's National Research and Development Strategy: A review'. *Science Technology and Society*, 9(2), pp 273–294.

Kaplan, D. 2008. 'Science and technology policy in South Africa: Past performance and proposals for the future'. *Science Technology and Society*, 13(1), pp 95–122.

Kaplan, D. 2009. 'Intellectual property rights and innovation in South Africa. A framework'. *The Economics of Intellectual Property in South Africa*. Geneva: World Intellectual Property Organization.

Kim, L. 1980. 'Stages of development of industrial technology in a developing country: A model'. *Research Policy*, 9(3), pp 254–277.

Kim, L. 1997. *Imitation to Innovation: The Dynamics of Korea's Technological Learning*. Boston: Harvard Business School Press.

Krishna, VV, Patra, SK & Bhattacharya, S. 2012. 'Internationalisation of R&D and global nature of innovation: Emerging trends in India. *Science Technology & Society*, 17(2), pp 165–199.

Kumar, N. 1990. *Multinational Enterprises in India: Industrial Distribution, Characteristics, and Performance*. London and New York: Routledge.

Lall, S. 1990. *Building Industrial Competitiveness in Developing Countries*. Paris: Organisation for Economic Co-operation and Development.

Lall, S. 1992. 'Technological capabilities and industrialisation'. *World Development*, 20(2), pp 165–186.

Lall, S. 1998. 'Technological capabilities in emerging Asia'. *Oxford Development Studies*, 26(2), pp 213–243.

Manzini, ST. 2015. 'Measurement of innovation in South Africa: An analysis of survey metrics and recommendations'. *South African Journal of Science*, 111(11/12), pp 1–8.

Morrison, A, Pietrobelli, C & Rabellotti, R. 2008. 'Global value chains and technological capabilities: A framework to study learning and innovation in developing countries. *Oxford Development Studies,* 36(1), pp 39–58.

Muchie, M. 2003. 'Re-thinking Africa's development through the national innovation system'. In M Muchie, P Gammeltoft & BA Lundvall, eds. *Putting Africa First: The Making of African Innovation System*. Aalborg, Denmark: Aalborg University Press.

Mytelka, LK. 2003. 'The dynamics of catching up: The relevance of an innovation system approach in Africa'. In M Muchie, P Gammeltoft & BA Lundvall, eds. *Putting Africa First: The Making of African Innovation System*. Aalborg, Denmark: Aalborg University Press.

Nagaoka, S, Motohashi, K & Goto, A. 2011. 'Patent statistics as an innovation indicator'. In BH Hall & N Rosenberg, eds. *Handbook of the Economics of Innovation*, pp 1083–1127. Amsterdam, The Netherlands: Elsevier.

Nagaratnam, A. 2001. 'Biotechnology in India: Current scene'. *Defense Science Journal*, 51(4), pp 401–408.

Nelson, RR. 1987. 'Roles of government in a mixed economy'. *Journal of Policy Analysis and Management*, 6(4), pp 541–566.

Nelson, RR & Winter, SG. 1982. *An Evolutionary Theory of Economic Change*. Massachusetts, USA: Harvard University Press.

OECD. 2007. 'Reviews of innovation policy: South Africa'. Paris: Organization for Economic Co-operation and Development.

OECD. 2008. *Compendium of Patent Statistics*. Paris: OECD.

OECD. 2011. *OECD Patent Database*. Paris: OECD

Patel, P & Pavitt, K. 1995. 'Patterns of technological change'. In P Stoneman, ed. *Handbook of the Economics of Innovation and Technological Change*. Hoboken, New Jersey: Wiley-Blackwell.

Patra, SK. 2014. 'Knowledge production by Indian biotechnology parks'. *Asian Biotechnology and Development Review*, 16(1), pp 35–53.

Patra, SK & Chand, P. 2005. 'Biotechnology research profile of India'. *Scientometrics*, 63(3), pp 583–597.

Patra, SK & Krishna, VV. 2015. 'Globalization of R&D and open innovation: Linkages of foreign R&D centers in India'. *Journal of Open Innovation: Technology, Market and Complexity,* 1(1). doi:10.1186/s40852-015-0008-6.

Pavitt, K. 1985. 'Patent statistics as indicators of innovative activities: Possibilities and problems'. *Scientometrics,* 7(1-2), pp 77–99.

Peet, NP. 2005. 'Biotechnology in India'. *Drug Discovery Today*, 10(17), pp 1137–1139.

Reid, SE & Ramani, SV. 2012. 'The harnessing of biotechnology in India: Which roads to travel?' *Technological Forecasting & Social Change*, 79(4), pp 648–664.

Satyanand, PN. 2007. 'Regions: Asia: Chemical attraction'. FDI Intelligence. Available at: www.fdiintelligence.com/Locations/Asia-Pacific/Chemical-attraction?ct=true (accessed on January 2018).

Smith, K. 2006. 'Measuring innovation'. In J Fagerberg & DC Mowery, eds. *The Oxford Handbook of Innovation*, pp 148–177. Oxford: Oxford University Press.

Solow, RM. 1956. 'A contribution to the theory of economic growth'. *The Quarterly Journal of Economics*, 70(1), pp 65–94.

Statement on Industry Policy. 1991. Available at: www.dipp.nic.in/sites/default/files/chap001_0_0.pdf (accessed on January 2018).

Uctu, R & Essop, H. 2013. 'The role of government in developing the biotechnology industry: A South African perspective'. *Prometheus,* 31(1), pp 21–33.

SEVEN

The proliferation of stem cell research and therapy in South Africa and India

A comparative study

SHASHANK S TIWARI

INTRODUCTION

RESEARCH AND INNOVATION IN STEM CELL technology has become a major focus in both developed and developing economies. Stem cells' capacity to develop into any kind of cells, tissues and organs make them an important tool in healthcare and biotechnology. Countries across the world have started many programmes and made significant investments in stem cells, anticipating cures for various presently incurable diseases and new contributions to understanding basic biology. At nation-state level, there is excitement that stem cells will be able to boost the biotechnology and pharmaceutical industries in the future. It is seen as a novel form of biomedicine, which brings both health and economic benefits. The area of stem cell treatment is estimated to be a billion dollar industry, and as a result stem cell research is becoming increasingly competitive worldwide (Tiwari & Desai, 2011; Lee et al, 2014).

A recent report suggests that the USA dominates the stem cell field in terms of number of publications. However, in terms of relative activity levels (i.e. country output levels to global activity level) Singapore is at the top followed by Italy, the USA, Japan and Israel (Barefoot et al, 2013). In addition to research activities, the stem cell clinical industry is also on the rise worldwide (Berger et al, 2016). BRICS countries (Brazil, the Russian Federation, India, China and South Africa) have also taken initiatives to capitalise on the underlying promise of stem cells to address their health needs.

Among BRICS nations, China could be considered as a leader in the area of stem cell research given that it is second on the list after the USA of countries producing the highest volume of publications. Stem cell activities in India, South Africa, Russia and Brazil have also increased in recent years (Cohen & Cohen, 2010; McMahon et al, 2010; Ballo et al, 2013; Tiwari, 2013). These five nations are viewed as having the potential to become the world's fastest growing economies. However, they are facing various challenges in providing basic healthcare facilities to their citizens. For BRICS nations, the promise of stem cells being a regenerative medicine could be an answer to many health challenges, more specifically for chronic and degenerative diseases.

Against this backdrop, this chapter aims to analyse the proliferation of stem cell research in South Africa and India. Both countries have recently come up with new biotechnology strategies and both face more or less similar challenges in the area of stem cells especially with knowledge production and regulations. The chapter analyses the stem cell development in South Africa and India using the national innovation system (NIS) as an analytical framework.

The major thrust of this study is to examine:

- knowledge production;
- the role of key actors and linkages including institutions, i.e. rules and regulations; and
- various other factors which are crucial for stem cell innovation in both countries.

ANALYTICAL FRAMEWORK: NATIONAL INNOVATION SYSTEM

The national innovation system (NIS) perspective emerged in the mid-1980s to develop a system perspective on innovation (Freeman, 1987; Lundvall, 1992; Nelson, 1993). NIS is used to examine the innovative ability of a nation-state, including the identification of country-specific similarities and differences with regard to innovation. It provides an analytical tool for the analysis of knowledge production and its diffusion, including the use of new knowledge linked to a particular technological development within a nation-state. There are two types of knowledge discussed in innovation systems: a) the codified scientific and technical knowledge and b) learning and experience based technical know-how, i.e. tacit knowledge (Jensen et al, 2007).

The NIS approach puts greater emphasis on the role played by firms. It also considers the importance of other actors such as universities, research institutes and government for a successful technological innovation (Lundvall, 1992). In innovation systems, the role of institutions is equally important. 'Insitution' is defined in various ways. The most accepted definition of institutions is the basic rules of the game. However, it is also used for players or actors (Lundvall 1992; Nelson & Rosenberg 1993). This suggests that there is no consistency in defining institutions in the NIS approach.

The scholarship surrounding innovation systems suggest that the linkages among key actors is necessary for a successful technological innovation. Innovative performance depends on how these actors interact with each other and this can be in various forms and at different levels. Niosi et al (1993: 210) stated:

> *interaction among these units may be technical, commercial, legal, social, and financial, in as much as the goal of the interaction is the development, protection, financing, or regulation of new science and technology.*

Interaction is crucial for the learning process; it helps in diffusion of knowledge and hence innovation. Edquist (1997) opined that the development of innovation, its diffusion and use is highly influenced by major economic, social, political, organisational and other factors.

This chapter aims to study the stem cell innovation in South Africa and India. Many studies in social sciences have highlighted the profound ethical and social factors that might affect the proliferation of stem cell innovation. The highly cited Hwang[20] affair suggests that these factors are crucial during the innovation process and that a robust regulatory framework is required along with a state-of-the-art scientific infrastructure for a successful innovation, i.e. 'co-production' of science and institutions is required (Jasanoff, 2004). While examining stem cell innovation in the USA, Salter and Salter (2010) argued that stem cell science is highly complex in terms of knowledge production and its translation. At various stages of innovation there is a tension between science, society and the market. For instance, 'the science may prove to be inadequate, society unsympathetic or the market uninterested' (Salter & Salter, 2010: 88). It can be argued, therefore, that in the case of those areas of science and technology, which have profound ethical and social issues, there is a need to have a robust framework for a successful innovation. This chapter offers an analysis of the country-specific innovation frameworks.

SOUTH AFRICA'S NATIONAL INNOVATION SYSTEM

In South Africa, the 'national innovation system', as a concept, was first introduced in the 1996 White Paper on science and technology. The paper conceptualised a national innovation system as 'a means by which a country seeks to create, acquire, diffuse and put into practice new knowledge that will help the country and its people achieve their individual and collective goals' (Department of Arts, Culture, Science and

20 According to *Nature* magazine, the South Korean scientist, Woo Suk Hwang, once feted for creating human stem cell lines used in cloned embryos, admitted to having falsified his data, and was sentenced to imprisonment for embezzlement and bioethical violations ('Woo Suk Hwang convicted, but not of fraud' by David Cyranoski, *Nature* 461, 1181 (2009). Available online at www.nature.com/news/2009/091026/full/4611181a.html.

Technology, 1996). The Council for Scientific and Industrial Research (CSIR) defines NIS as a network of players in a country that interact to constitute the country's innovation system. The CSIR has given more importance to public sector research. It is believed that there is a greater socio-economic return to be had from public sector research and that this is more extensively applied in nature and over the long term. The CSIR emphasises the importance of government funding. The 1996 White Paper argued that 'the development and application of science and technology within a national system of innovation in South Africa will be central to the success of the Growth and Development Stragey (GDS) of the Government as it seeks to address the needs of all South Africans' (Department of Arts, Culture, Science and Technology, 1996). The White Paper was followed by the National Biotechnology Strategy (2001) which put emphasis on science-based innovation in health and agriculture. In 2008, the Department of Science and Technology proposed the Ten-Year Innovation Plan (2008–2018) to transform South Africa into a knowledge-based economy. The plan seeks to develop the South African bio-economy from, as the government puts it, 'Farmer to Pharma'. South Africa updated its biotechnology strategy in 2013 by putting equal emphasis on social sciences, engineering and information technology. The 2013 bio-economy strategy plans for innovation in biotechnology to contribute significantly to South Africa's GDP by 2030. The strategy laid out a vision for a healthy South Africa, based on strengthened local research, development and innovation capabilities. This strategy paid attention to ethical and social issues that have been raised with the emergence of modern biotechnology. This document emphasises that,

> *The regulatory landscape should address the ethical implications of all innovations, while taking care not to stifle research and innovation. The country should strike a healthy balance between recognising the potential benefits of biotechnology and ensuring that research is ethically conducted. This will require constantly evolving ethical and regulatory frameworks* (Department of Science and Technology, 2013: 21).

This strategy clearly lays out plans for the promises of stem cell innovation to be explored.

INDIA'S NATIONAL INNOVATION SYSTEM

India's national innovation system, like South Africa's, is also dominated by government institutes. In 2010, the government of India created the National Innovation Council and declared 2010–20 the 'Decade of Innovation'. The 2013 Science, Technology and Innovation Policy further emphasises science-led innovation in the Indian context. This innovation policy document advocates for an academia–research–industry partnerships in R&D and aims to have a greater role played by the private sector. The Department of Biotechnology, the Department of Science and Technology and the Council for Scientific and Industrial Research are playing a major role in science-led innovation in India through funding and policy formulation. In the new biotechnology strategy 2015–2020, the government of India puts greater emphasis on understanding the basic biology of stem cells including stem cell-based cell therapies. Additionally, this strategy advocates for a transparent and efficient regulatory system to ensure the safety of consumers and the environment.

METHODS

To analyse the current status of stem cell research in South Africa and India, including the role of key actors, linkages and regulations in this research, a qualitative study of documents (scientific literature, policy reports and news) and interviews with key actors were undertaken. The interviews in South Africa were conducted during 2014–15 via telephone with leading scientists and bio-legal experts. The data on Indian stem cell research is largely based on my own PhD dissertation

(2013) at the University of Nottingham (UK)[21] and it is further updated with recent developments.

STEM CELL RESEARCH IN SOUTH AFRICA AND INDIA

The proliferation of stem cell research and therapy in India can be traced back to the beginning of the 1980s, in the wake of India's first successful allogenic bone marrow transplantation (a type of stem cell transplantation) on 20 March 1983 (Chakraborty et al, 2009). In South Africa it is argued that stem cell transplantation was first introduced in 1970 (Wood et al, 2009). This suggests that South Africa was well ahead of India in introducing stem cell-related activities.

Government supports stem cell programmes in both countries, especially basic research. In their respective biotech strategies, there is an emphasis on the need to use knowledge of biotechnology to address local health needs and stem cell-based therapies are viewed as one of the important tools available to tackle the growing disease burden. However, the legitimate question remains with respect to their innovative capacity and whether during the innovation process crucial social and ethical factors are taken into account.

STEM CELL KNOWLEDGE PRODUCTION AND LINKAGES IN SOUTH AFRICA

Although stem cell activities in South Africa are at a nascent stage when compared to other BRICS countries, there is a clear indication that South Africa has the potential to conduct state-of-the-art research in stem cells. In recent years, a few universities have started stem cell research programmes. South Africa has also become a favourable destination for stem cell-based treatments, a phenomenon widely

21 Tiwari, S. 2013. 'The ethics and governance of stem cell clinical research in India'. Available at: www.eprints.nottingham.ac.uk/14585/1/602957.pdf.

popularised as 'stem cell tourism'. It has attracted many local as well as foreign patients towards hospitals and companies who claim to cure various conditions for which existing treatments have been exhausted. In addition, it is also worth noting the rise of private cord blood banking companies.

Publicly funded research institutes and universities in South Africa are playing a leading role in the production of knowledge about stem cells. The major focus is on adult stem cells and induced pluripotent stem cells (iPS). The Stem Cell Research Initiative (SCI) at the University of Cape Town is working on a 'disease in a dish' model as informed by a scientist based at SCI. The aim of the 'disease in a dish' model is to create iPS cells. The creation of iPS cells was first reported in South Africa in the year 2012 by a group of scientists from the Council for Scientific and Industrial Research (CSIR). It was heralded as a breakthrough as a very few developed countries, including India, have been able to carry out this technology. In addition to iPS, the scientists at SCI are focusing on inherited ataxias, myasthenia gravis, and retinal and corneal degeneration. Currently, the institute is working in collaboration with many national and international institutes.

Scientists at the University of Pretoria are engaged in adult stem cell research. A scientist who is conducting research on adult stem cells stated that 'We are working on haematopoietic stem cells and mesenchymal stem cells' (Scientist 2).

The long-term goal at the University of Pretoria is to develop the capability of somatic cell nuclear translation. The aim is to address the high disease burden of both communicable and non-communicable diseases. Scientist 2 stated that:

> *We have a very high disease burden, both communicable and non-communicable diseases. We look at communicable diseases, the highest burden of HIV and TB with a contribution also from malaria. There are some experimental treatments that are being received both for HIV and for TB. But I think for a long time ... that we've got non-communicable diseases such as myocardial infarction, stroke, high blood pressure, leukaemia, some would say going on probably in the area of diabetes.*

To address the high disease burden as highlighted above, the University of Pretoria has been awarded a Medical Research Council Extramural Research Unit for Stem Cell Research and Therapy that started working in 2015. This can be viewed as a serious effort towards stem cell innovation in the country.

At the University of Johannesburg, the main focus is on the use of low intensity laser irradiation in stimulating derived stem cell proliferation. Similarly, scientists at the Biomedical Biotechnology Research Unit (BBRU), at Rhodes University are involved in cancer stem cell research. The stem cell research laboratory at the University of KwaZulu-Natal focuses on the role of cytokines in regulating stem cell quiescence, proliferation, migration and differentiation. In addition to these universities, CSIR, as highlighted above, conducts research on iPS cells. The CSIR has established strong linkages with many universities for conducting basic research. For instance, the CSIR is working with the Stem Cell Research Initiative at the University of Cape Town in iPSC-based research. The MRC Extramural Stem Cell Unit at the University of Pretoria works with the Reproductive Biology Laboratory at the Steve Biko Academic Hospital, University of Pretoria for a somatic cell nuclear transfer programme. This unit has also collaborated with countries such as the USA, France and Switzerland.

Although there are indications of linkages between academic institutes and research-oriented hospitals in South Africa, there is no evidence of linkages between universities and private firms. The reason might be that in the area of stem cells, not many firms are interested in research and development. Rather they are more interested in providing stem cell-based services such as cord blood banking, though a few are involved in stem cell therapies.

UMBILICAL CORD BLOOD BANKING IN SOUTH AFRICA

Lazaron Biotechnologies, founded in 2005, is considered the first cord blood banking company in South Africa. However, since 2011, it has been administered by a multinational cord blood banking company, Cryo-Save. Netcells Biosciences, based in Johannesburg, is characterised as Africa's largest cord blood bank (Jackson & Pepper, 2013). Switzerland-based Salveo Swiss Biotechnology has also started banking services in South Africa. It is a joint venture of JSE-listed Ecsponent Limited and the international biotechnology group, Esperite.

The cost of the banking facility varies from company to company and it also depends on how long expectant parents opt for storage. Netcells, for example, charges R18 350 for five-year storage and R19 600 for 10 years. It is worth highlighting that there is no public cord blood bank in South Africa. There are many advocates for public cord blood banking against the backdrop of genetic diversity in South Africa (Mellet et al, 2015). It is argued that the South African Bone Marrow Registry does not represent the majority Black African population, rather it is dominated by the Caucasian population (Mellet et al 2015). The absence of public cord blood banking could lead to health inequality in South Africa. It would be very hard for a large section of society to make use of the benefits of stem cell transplantations.

STEM CELL CLINICAL APPLICATIONS IN SOUTH AFRICA

In South Africa, stem cell-based therapies are being offered mostly by private hospitals using adult stem cells. The Netcare Pretoria hospitals appears to be the leading hospital in stem cell therapy which claims to perform 80 stem cell transplants a year, using haematopoietic stem cells (Dugmore, 2014). Melomed Bellville Private Hospital, Cape Town is also providing stem cell therapy, using adult stem cells and somatic cell nuclear transfer (Du Toit & Liebenberg, 2014). Various hospitals and individual clinicians have not only attracted domestic patients but international patients as well. However, there are serious concerns about the safety and efficacy of the treatments as many treatment procedures are not backed by enough scientific evidence in the form of clinical trials and largely seen as unproven therapies. There is concern that stem cell clinical activities in South Africa are proliferating in a legislative vacuum according to one of the scientists I interviewed (Scientist 2). In 2012, it was reported that a clinician and his team treated a patients for spinal cord injuries using therapeutic stem cell cloning without having prior ethical approval. The said clinician also did not secure an approval from the Medical Control Council (Slabbert et al, 2015). This suggests that it is necessary to examine the current regulatory framework in South Africa.

STEM CELL REGULATION IN SOUTH AFRICA

The regulation of stem cell innovation in South Africa appears to be highly complex given that directly or indirectly many Acts are associated with the regulation of stem cell activities in the country. Though primarily the National Health Act (NHA) 2003 deals with the regulation of cell-based therapies, including stem cells and other important health issues, the other national Acts such as Medicines and Related Substances Act, the Consumer Protection Act, the Children's Act including the Inquest Act are also applicable in regulating stem cell-based therapy (Pepper & Slabbert, 2015). The NHA came into force on 2 May 2005. However, the main provisions dealing with human tissue, i.e. Chapter 8, was enacted only on 2 March 2012.

Chapter 8 of this Act aims to 'regulate of use of blood, blood products, tissue and gametes in humans'. The fact that these provisions of Chapter 8 were only enacted by 2 March 2012 suggests that stem cell activities in South Africa were proceeding in a regulatory vacuum for many years. However, in the current scenario it appears that in South Africa, the government has now made an effort to address unregulated stem cell therapy.

However, the question remains whether after the commencement of Chapter 8 the vacuum in stem cell regulation has been addressed. A scientist opined that:

> *Chapter 8 of the National Act is enacted which basically deals with human biological material and the control of body and*

> *tissues and things like that. There are regulations relating to stem cell banks but with regard to actual therapies no regulations came out. The regulations that were enacted have their own sets of problems I think it was a bit rushed* (Scientist 2).

The above statement infers that even after the enactment of Chapter 8, the situation remains the same, especially in the case of stem cell therapies. Scientist 2 argued that there are no regulations with regard to stem cell-based therapies. However, the NHA and Medicine Council Control's (MCC) guidelines including other Acts suggest that there are enough regulations for stem cell. As per MCC guidelines, stem cell intervention in the human body is categorised as a biological medicine and according to the Medicines and Related Substances Control Act (MRSCA), any biological medicine requires registration with the MCC. In addition, a particular biological medicine should have proven efficacy, safety and quality (Botes & Alessandrini, 2015). The intervention of stem cells in the human body is, therefore, subject to the approval of the MCC. However, it is noted that at many occasions a few clinicians have provided stem cell therapies without having approval from regulatory bodies such as the MCC.

This suggests that there is a problem with the implementation of laws and guidelines in South Africa. A scientist informed me that in South Africa 'it (stem cell intervention) is administered without going through a clinical trial phase and the MCC should intervene as per their jurisdiction' (Scientist 1). She further stated, 'so we have a set of guidelines but who enforces it, who makes sure, who checks up that it is being carried out in the correct way' (Scientist 1).

However, recently the MCC has taken action against an illegal stem cell treatment contract (for treatment of knee patients) between the Free State Department of Health and the stem cell company ReGenesis Biotechnologies. According to a media report, the department would have to pay R30 000 per client to the company with the assurance that it would supply 1000 patients per year. It appears that the initial contract was for three years. The MCC clarified that this particular treatment for knee patients is not a proven therapy and that the MCC has not given approval to conduct this type of treatment. Finally, the Free State

Department of Health cancelled the contract (Thom & Low, 2016). It seems, therefore, that the MCC has started enforcing its jurisdiction to prevent the use of experimental stem cell therapy.

STEM CELL KNOWLEDGE PRODUCTION AND LINKAGES IN INDIA

In India, as in South Africa, basic research in stem cells is being carried out largely by government-funded research laboratories. These research laboratories are supported by the different government agencies such as the Department of Biotechnology (DBT), the Council for Scientific and Industrial Research (CSIR), the Indian Council of Medical Research (ICMR), the Department of Science & Technology (DST), etc.

The focus is mainly on adult stem cells. Different institutions are engaged in different areas of adult stem cell research. Only a few research laboratories, compared to those dealing with adult stem cells, are focusing in the area of embryonic stem cells (Tiwari & Desai, 2011). Some research laboratories are also researching umbilical cord blood cells.

Some centres in India have developed state-of-the-art stem cell technology. For example, the Centre for Stem Cell Research (CSCR), Vellore has succeeded in the creation of Induced Pluripotent stem cells (iPS). In media commentary, this is portrayed as ground-breaking research.[22] India has recently established the Institute for Stem Cell Biology and Regenerative Medicine (inStem), with an investment of US$50 million in Bangalore. It is viewed as a major boost for stem cell R&D in India (Sachitanand, 2009).

A multi-disciplinary research institute, Jawaharlal Nehru Centre for

22 See, for example, articles in the *Times* of India newspaper: www.timesofindia.indiatimes.com/home/science/After-funding-for-brain-research-Infosyss-Kris-Gopalakrishnan-now-turns-eye-on-stem-cell-research/articleshow/50655903.cms and on the Business Insider website: www.businessinsider.in/Stem-CellResearch-In-India-Surges-Ahead/articleshow/40278445.cms

Advance Scientific Research (JNCASR), which is also in Bangalore, has derived two human embryonic stem cell (hESC) lines from discarded embryos. These cell lines are available through a UK stem cell bank to carry out further research. The Centre for Cellular and Molecular Biology (CCMB), one of the constituent national laboratories of CSIR, based in Hyderabad, has established a stem cell facility centre in association with DST and Nizam's Institute of Medical Sciences (NIMS), Hyderabad. The facility centre aims to work in both basic and applied research. For applied research, NIMS will provide patients to the CCMB for applied research (CCMB, 2009; Tiwari & Desai, 2011).

In contrast to public research laboratories, hospitals in India are engaged in both basic as well as clinical research. The majority of hospitals, specifically private hospitals, offer direct therapy to patients; however, a few public hospitals are also engaged in the therapeutic applications of stem cells. Public hospitals are mostly involved in basic/clinical research in stem cells. The All India Institute of Medical Sciences (AIIMS), New Delhi, which is India's premier research-intensive public hospital, set up a stem cell facility for this purpose in 2005. During a fieldwork visit, a clinician at this centre informed me that:

> *The stem cell facility at AIIMS was established for the purpose of doing basic and clinical research and we are trying to see its [stem cell] role in degenerative areas like cardiac, neurological, epithelial surface, eye and skin* (Clinician 7).

Similar to the AIIMS, other hospitals/medical institutes such as the Post Graduate Institute of Medical Sciences (PGI), Chandigarh, the Sri Venkateswara Institute of Medical Sciences (SVIMS), Tirupati and the Sanjay Gandhi Post Graduate Institute of Medical Sciences (SGPGI), Lucknow are conducting basic and clinical research in the different areas of stem cell therapy.

In addition to these, a not-for-profit hospital, the L.V. Prasad Eye Institute (LVPEI), Hyderabad is also active in research and development of stem cell-based therapy. LVPEI is conducting both basic and clinical research in the area of stem cells, specifically in limbal stem cells.

Sankara Nethralaya, an eye hospital based in Chennai, is conducting research in the area of corneal and retinal stem cells. Recently, it has developed synthetic gel to grow stem cell corneas and also obtained an international patent for this technology (Narayan, 2010). It can be seen as an indicator of India's growing scientific strength in the field of stem cells, along with publications.

In contrast to research laboratories and hospitals, a few firms are doing basic research along with clinical research. The Reliance Life Science Pvt. Ltd (RLS), Mumbai is conducting research in embryonic and adult, as well as cord blood, stem cells. RLS is actively involved in developing stem cell technology in India, as per information provided by the company's representative during the field visit (Firm representative 6). Under the Regenerative Medicine initiative, RLS has set up several groups to work in the areas of embryonic stem cells, ocular stem cells, haematopoietic stem cells and skin and tissue engineering.

Stempeutics Research Pvt. Ltd is another important firm in India primarily engaged in developing stem cell-based products. Stempeutics mainly focuses on adult stem cell research, although they are also interested in embryonic stem cell research but it is a long-term goal only, according to a representative of Stempeutics who informed me that:

> *We do research on embryonic stem cells on a small scale considering the potential for embryonic stem cells in the future but it is not our short-term or mid-term goal; this is our long-term goal. As for our short-term and mid-term goals, we are focusing on adult stem cells* (Firm representative 4).

In adult stem cells, Stempeutics is mainly involved in mesenchymal stem cells (MSCs) derived from sources such as bone marrow, adipose tissue, Wharton's jelly and dental pulp. Recently, Stempeutics has been granted a process patent for its stem cell based drugs Stempeucel. It is argued that this drug will be useful for treatment of Critical Limb Ischemia (Subbu 2015).

The Nichi-In Centre for Regenerative Medicine (NCRM) is also seen as one of the leading firms in the stem cell sector in India. It is an Indo-Japanese joint-venture firm mainly focusing on autologous adult stem cells; more specifically limbal, haematopoietic, mesenchymal liver and corneal endothelial precursor stem cells. It has linkages with different hospitals in the country for research and clinical application of stem cells (Tiwari & Desai, 2011). However, the work of NCRM as a stem cell service provider was a source of criticism in 2008 because of its nexus with the Life Line Hospital, Chennai, which offers unproven therapy to patients (Pandya, 2008).

In India, to translate basic research at a clinical level, government and private players have taken major initiatives. The mandate behind the establishment of inStem is very much in this direction. inStem works in collaboration with many national and international institutes based in the USA and Japan. This research institute has established a linkage with the clinicians of CMC, Vellore. Additionally, the DBT has started city cluster programmes to established linkages among different government institutes including firms. For basic research, LVPEI is working with many national and international institutes. In association with the University of Sheffield, UK, LVPEI has developed biocompatible materials for stem cell transplantation. The aim of this collaborative project is to develop affordable innovative healthcare products.

STEM CELL CLINICAL APPLICATIONS IN INDIA

In India, significant numbers of public and private hospitals are involved in the clinical application of stem cells. They are offering various kinds of stem cell-based therapies to patients for a wide range of incurable diseases using adult stem cells. Only Nutech Mediworld is offering embryonic stem cell treatments. Stem cell-based treatments have attracted a significant number of local and international patients to various hospitals in the country (Cohen & Cohen, 2010). However, the therapeutic applications of stem cells in India have raised several ethical issues in the absence of regulatory approval and clinical trials.

AIIMS has done a few stem cell transplantations for some

conditions over the years. The first use of stem cells among 35 cardiac patients by the clinicians at AIIMS was reported in the media and in science magazines, like *Biospectrum*, in 2005. The stem cell injection was given to these patients between February 2003 and January 2005 (*Biospectrum*, 2005). However, these stem cell transplantations were criticised on ethical and safety grounds since AIIMS failed to provide any proof of the ethical clarity or of experimental and clinical studies prior to applying this therapy to the patients (Pandya, 2008). There was also a report in the *Indian Express* newspaper (Krishnan, 2009) that AIIMS did not fully inform the patients about the pros and cons of stem cell treatments. Following the widespread criticisms, AIIMS made an announcement about the abandonment of this kind of treatment. It appears that AIIMS is now advancing in this area cautiously as it has recently dropped two stem cell trials related to muscular dystrophy and motor neuron disease because of poor patient response (Krishnan, 2009).

LVPEI has treated over 700 patients (Mallikarjun 2012). It is worth highlighting here that many patients were treated free of charge (Lander et al, 2008; Vemuganti & Sangwan, 2010; Tiwari & Desai, 2011). However, in 2006, the stem cell transplantation procedure at LVPEI came under scrutiny when a team of international ophthalmologists criticised the existing stem transplantation procedures, in which human and/or animal materials were being used for limbal stem cell growth (Schwab et al, 2006). These ophthalmologists had reviewed all the clinical trials worldwide between the period 1 July 1996 and 30 June 2005, including LVPEI clinical trials. The use of bioengineered ocular surface tissue in transplantation procedures was examined during this review. It was observed that the 'current investigational protocols rely on the use of animal and/or donor human tissue products and thus carry the potential to induce xenogenic microchimerism in recipients or disease transmission through contamination with bacteria, viruses, or other infectious agents, such as those responsible for transmissible spongiform encephalopathy' (Schwab et al, 2006: 1 734). In response to this report, a clinician at LVPEI admitted the 'potential risk'. However, he denied any adverse effects of stem cell transplantations in his patients (*Gulf Times*, 2007).

Nutech Mediworld has attracted great attention on national and international platforms for its embryonic stem cell treatments. This is a private stem cell clinic run by an obstetrician, Dr Geeta Shroff, in New Delhi. She claims to have treated over 1 400 patients from different countries including India. It is reported that she has been granted patents in many countries, including the USA, Japan, Australia, South Korea, Australia and New Zealand (*Business Standard,* 2016). However, Dr Shroff's work is not accepted by many scientists and clinicians, including policy-makers, in the absence of clinical trials. In December 2010, there were reports in the media that Drug Controller General of India (DCGI) had asked Dr Shroff to explain her claims to have successfully treated patients and to show whether she has an approval from any regulatory body (Pandeya, 2010). Similar action was also taken from the government of India in 2006 when an inquiry was set up against Dr Shroff's stem cell treatments. According to *The Hindu* newspaper, there was promise from the Ministry of Health that strict action would be taken once the inquiry report came out. However, Dr Shroff still continues with her work (Cohen & Cohen, 2010; *The Hindu*, 2006).

The Chaitanya Stem Cell Centre is part of Chaitanya Hospital, located in Pune. This centre offers therapy for a wide range of diseases including spinal cord injury, diabetes, stroke, autism, liver diseases and motor neuron diseases. However, the DCGI has asked this centre to stop stem cell treatments in the name of clinical trials. It was found that this centre is involved in unethical clinical stem cell clinical trial malpractices (Shankar 2014). The website of Chaitanya Stem Cell Centre is fully loaded with patients' testimonials, which is a kind of 'informal advertisement' and violates the medical code of ethics as per the Medical Council of India.

The Fortis Hospital in Delhi offers stem cell treatments for arthritis, bone regeneration, cartilage regeneration, non-healing wounds and chronic pain. In the beginning of 2011, Fortis Hospital attracted attention from the media because of the treatment of an Iraqi national using stem cell injection. However, the ICMR raised concern over this unproven therapy offered by the Fortis Hospital since it is still an experimental therapy and has not been authorised in India (Chatterjee, 2011).

A few banking firms also took the opportunity offered by the clinical application of stem cells. For example, International Stem Cell Services Ltd (ISSL) is a stem cell banking company located in Bangalore. In recent years, it started offering various stem cell-based treatments for disease conditions such as Buerger's disease (critical limb ischemia), chronic liver and kidney failure, osteoarthritis, muscular dystrophy, myocardial infarction and spinal cord injury. ISSL has also conducted clinical trials for some of the abovementioned disease conditions as per information provided by the representative of firms interviewed during fieldwork (Firm representative 5).

UMBILICAL CORD BLOOD BANKING IN INDIA

Compared to South Africa, there are greater numbers of private cord blood banking firms in India. Currently, there are 14 licensed/registered private UCB banks in operation as per information provided by the health ministry (Ministry of Health and Family Welfare, 2015). However, unlike in South Africa where there is no public cord blood bank, in India, there are four public banks, out of which one is run by the government based at the School of Tropical Medicine, Kolkata.

Life Cell International in India is viewed as one of the leading cord blood banking firms; it started functioning in the year 2004 in collaboration with CRYO-CELL International Inc. and is based in Chennai. The transplant centre of the company is set up at Sri Ramchandra Medical College, Chennai. Cryobanks International India is a joint venture between Cryobanks International USA and RJ Corp India, based in Gurgaon near Delhi. It was established in 2006. It has also shown interest in the development of stem cell therapies for cardiac disease and diabetes. CordLife Sciences India Pvt. Ltd is located in Kolkata. It was launched in 2006. It is part of the Cordlife group, which is considered to be Australasia's largest network of private cord blood banks. Cryo-Save India was established in 2009. It is based in Bangalore. It is a subsidiary of Cryo-Save, Netherlands. The company is keen in expanding its business into rural areas (Tiwari, 2013). However, the affordability could be a key issue for rural areas as the current storage cost is in the range of INR55 000 (R10 795) to INR75 000 (R14 720).

Out of four public banks, the Jeevan Blood Bank and Research Centre, which is located in Chennai, could be viewed as the first purely public bank in India. It is a not-for-profit blood bank established in 1995. It started its stem cell umbilical cord blood unit in 2010 as per the information provided during the field visit (Firm representative 1). Other private banks such as Reliance Life Sciences Mumbai and StemCyte India Gandhinagar also offer public cord blood banking services.

STEM CELL REGULATIONS IN INDIA

The regulatory environment in India, similar to South Africa, is also very complex because various Acts related to healthcare, including the Consumer Protection Act, are applicable in regulating stem cell activities. In addition to various acts, stem cell activities in India are directly governed by the ICMR-DBT guidelines, which were recently updated in 2013 (Tiwari & Raman, 2014). These guidelines discourage stem cell-based therapies, in contrast to the 2007 guidelines, as it omitted the word 'therapy' from the title of the updated guidelines. In the Foreword, the guidelines emphasise that,

> *stem cells are still not a part of standard care; hence there can be no guidelines for therapy until efficacy is proven. These guidelines are intended to cover only stem cell research, both basic and translational, and not therapy. It has been made clear in these guidelines that any stem cell use in patients, other than that for hematopoietic stem cell reconstitution for approved indications, is investigational at present. Accordingly, any stem cell use in patients must only be done within the purview of an approved and monitored clinical trial with the intent to advance science and medicine, and not offering it as therapy. In accordance with this stringent definition, every use of stem cells in patients outside an approved clinical trial shall be considered as malpractice* (ICMR-DBT, 2013).

The revised 2017 guidelines (ICMR-DBT, 2017) also reiterated a similar position (Tiwari et al 2017).

However, since these guidelines are not legally binding, clinicians appear to be free to offer experimental stem cell therapies to patients. It is worth mentioning that neither ICMR nor DBT are oversight authorities for regulating stem cell practices in India. The ICMR acts as an advisory body on healthcare issues and the DBT is primarily a funding agency to support biotechnology programmes including stem cell research. Stem cell clinical activities in India can only be governed through the Medical Council of India, which is the licensing authority for clinical practices, and Central Drugs Standard Control Organization (CDSCO), which oversees the use of drugs and clinical trials. In India, stem cell intervention still falls under the purview of medical practices. To date, however, the Medical Council of India has never taken any action against any clinicians.

To address unregulated stem cell therapies, the CDSCO had set up a committee to explore how the usage of stem cells can be regulated. One of the main recommendations of the committee is to amend the Drugs and Cosmetics Act, 1940 to include stem cells and other cell-based products as new drugs. As per recommendations of the committee, the Ministry of Health and Family Welfare has proposed to amend the Drugs and Cosmetics Act 1940 to include stem cells as drugs. The draft Drugs and Cosmetics (Amendment) Bill 2015 was released in January 2015 for the same reason.

CONCLUSION

Both South Africa and India have given priority to science and technological innovation aimed at addressing various societal problems including local health needs. Stem cell innovation is one of the important aspects of their biotechnology strategies. In both countries, research in stem cells is greatly supported by the government. Both countries have been able to develop iPS technology, which can address many ethical issues related to using human embryos in stem cell research. In India, in addition to government, private players are also involved in stem cell research, while in South Africa, it appears that firms are not interested in conducting basic research. In South Africa most of the private players are involved in either private cord blood banking business or offering experimental stem cell-based therapies to patients. South Africa is unable to develop any stem cell product, while in India a private firm has developed a stem cell product and obtained a patent in many countries. In contrast to South Africa, in India various linkages are visible both at national and international level, most importantly in a few cases firms have established linkages to universities. This is a key for knowledge diffusion. Though there is a sign of linkages in South Africa as well, it is comparatively limited and dominated by academic connections.

In terms of stem cell clinical applications, in both countries there are hospitals and clinics actively offering therapies for various disease conditions. Most of these hospitals are using adult stem cells and stem cells derived from cord blood for the treatments. However, in India, a clinician also uses stem cells derived from an embryo for various incurable diseases.

Stem cell-based therapies in both countries have raised significant social and ethical concerns as these therapies are being offered without any clinical trials. Both governments appear to be serious in curbing unproven stem cell activities as evident from updated guidelines and statutory measures. In contrast to India, regulations in South Africa appear to be more stringent given that the intervention of stem cells in patients is categorised as a biological medicine. However, India is still struggling to categorise stem cell clinical applications as a drug. In India, it is seen as a routine medical practice and for that there is no need for any registration to a regulatory authority. This leads to unregulated stem cell medical practices. However, there are many other healthcare legislation that can prevent experimental medical practices; the Indian regulatory authority has never taken any action to curb stem cell clinical activities. In contrast, recently South Africa's medical body has taken action against an illegal stem cell treatment contract. India is making an effort to categorised stem cells as drugs in order to have better regulation of stem cell therapies. However, it was noted that India has enough rules to regulate stem cell clinical activities; there is a need to work on implementing these oversight mechanisms (Tiwari & Raman, 2014).

Many experts in both South Africa and India are of the view that unethical and fraudulent activities in the stem cell sector stifle innovation (Meissner-Roloff & Pepper, 2013; Singh, 2013). This might be the reason behind putting emphasis on analysing the ethical and social implications of modern biotechnology in the respective government policies in South Africa and India. In innovation system framework, though it is acknowledged that the development of innovation, its diffusion and use is highly influenced by social factors among other things, the raised ethical issues linked to science and technological innovations and their impacts on innovations are not explicitly paid due attention. The discussion in this chapter suggests that for a successful stem cell innovation, South Africa and India need to have a better scientific infrastructure, including a robust regulatory framework to address unethical stem cell activities.

ACKNOWLEDGMENTS

The sections on Indian stem cells is part of a Wellcome Trust Studentship (grant number: WT087867MA) awarded to Shashank S Tiwari at the Institute for Science and Society, School of Sociology and Social Policy, University of Nottingham, Nottingham (UK). The Trust is not responsible for views expressed in this chapter. The author is grateful to Paul Martin, Sujatha Raman and Pranav Desai for support and guidance.

REFERENCES

Ballo, R, Greenberg, LJ & Kidson, SH. 2013.' A new class of stem cells in South Africa: iPS cells'. *SAMJ: South African Medical Journal*, 103(1), pp 16–17.

Barefoot, J, Doherty, K, Kemp, E, Blackburn, C, Sengoku, S, Van Servellen, A, Gavai, A & Karlsson, A. 2013. 'Stem cell research: Trends and perspectives on the evolving international landscape'. *EuroStemCell, Kyoto University's*

Institute for Integrated Cell-Material Sciences (WPI-iCeMS), and Elsevier'. Available at: www.elsevier.com/__data/assets/pdf_file/0005/53177/Stem-Cell-Report-Trends-and-Perspectives-on-the-Evolving-International-Landscape_Dec2013.pdf (accessed on 20 April 2015).

Berger, I, Ahmad, A, Bansal, A, Kapoor, T, Sipp, D & Rasko, JEJ. 2016. 'Global distribution of business marketing stem cell-based interventions'. *Cell Stem Cell*, 19(2), pp 158–162.

Botes, WM & Alessandrini, M. 2015. 'Legal implications of translational promises of unproven stem cell therapy'. *South African Journal of Bioethics and Law*, 8(2), pp 36–40.

Biospectrum. 2005. 'Cardiac patients get stem cell therapy at AIIMS'. Available at: www.biospectrumindia.com/news/73/1322/cardiac-patients-get-stem-cell-therapy-at-aiims.html (accessed on 20 June 2011).

Business Standard. 2016. 'Embryonic stem-cell therapy can treat incurable conditions: Researcher Geeta Shroff'. Available at: www.business-standard.com/article/news-ians/embryonic-stem-cell-therapy-can-treat-incurable-conditions-researcher-geeta-shroff-ians-interview-116012000532_1.html (accessed on 24 January 2017).

CCMB. 2009. 'Nizam team up for stem cell facility'. *The Financial Express*, 21 January.

Chakraborty, C, Shieh, P & Agoramoorthy, G. 2009. 'India's stem cell research and development perspectives'. *International Journal of Hematology*, 89, pp 406–408.

Chatterjee, P. 2011. 'Fortis uses stem cell shot, medical body calls it "illegal, unethical"'. *Indian Express,* 11 January. Available at: www.indianexpress.com/news/fortis-uses-stem-cell-shot-medical-body-calls-it-illegal-unethical/735865/0 (accessed on 20 February 2011).

Cohen, CB & Cohen, PJ. 2010. 'International stem cell tourism and the need for effective regulation, Part 1 – Stem cell tourism in Russia and India: Clinical research, innovative treatment, or unproven hype?' *Kennedy Institute of Ethics Journal,* 20(1), pp 27–49.

Department of Arts, Culture, Science and Technology. 1996. 'White paper on science and technology: Preparing for the 21st century'. Pretoria: DACST.

Department of Science and Technology. 2013. *The Bioeconomy Strategy.* Pretoria: Department of Science and Technology.

Du Toit, DF & Liebenberg, WA. 2014. 'Somatic-Cell Nuclear Transfer: Autologous Embryonic Intra-Spinal Stem Cell Transplant in a Chronic Complete Quadriplegic Patient. Neuro-Anatomical Outcome after One Year'. *Revista Argentina de Anatomía Clínica*, 6(1), pp 35–42.

Dugmore, H. 2014. 'Stem cell transplants are saving lives in SA', 8 October. Available at: www.biznews.com/columnists/heather-dugmore/2014/10/08/47395heather-dugmore-stem-cell-transplants-are-saving-lives-in-sa/ (accessed on 28 June 2015).

Edquist C. 1997. 'Systems of innovation approaches: Their emergence and characteristics'. In Edquist, C, ed. *Systems of Innovation: Technologies, Institutions and Organizations*. London: Pinter/Cassell.

Freeman, C. 1987. *Technology and Economic Performance: Lessons from Japan*. London: Pinter.

Gulf Times. 2007. 'Stem cell transplant for eye repair raises concern'. 18 January. Available at: www.gulftimes.com/site/topics/article.asp?cu_no=2&item_no=128025&version=1&template_id=40&parent_id=22 (accessed on 24 April 2010).

ICMR-DBT. 2013. *National Guidelines for Stem Cell Research*. Available at: www.dhr.gov.in/sites/default/files/NGSCR-2013_0_0.pdf (accessed in May 2016).

ICMR-DBT. 2017. *National Guidelines for Stem Cell Research*. Available at: www.icmr.nic.in/guidelines/Guidelines_for_stem_cell_research_2017.pdf (accessed in December 2017).

Jackson, CS & Pepper, MS. 2013. 'Opportunities and barriers to establishing a cell therapy programme in South Africa'. *Stem Cell Research & Therapy*, 4(3), p 54. doi.org/10.1186/scrt204.

Jasanoff, S. ed. 2004. *States of Knowledge: The Co-Production of Science and Social Order*. London: Routledge.

Jensen, MB, Johnson, B, Lorenz, E & Lundvall, B-Å. 2007. 'Forms of knowledge and modes of innovation'. *Research Policy*, 36(5), pp 680–693.

Krishnan, V. 2009. 'AIIMS drops two stem cell trials due to poor patient response. *Indian Express,* 7 February. Available at: www.indianexpress.com/news/aiims-drops-two-stemcell-trials-due-to-poor/420224/ (accessed on 18 August 2011).

Lander, B, Thorsteinsdttir, H, Singer, PA & Daar, AS. 2008. 'Harnessing stem cells for health needs in India'. *Cell Stem Cell*, 3(1), pp 11–15.

Lee, S, Kwon, T, Chung, EK & Lee, JW. 2014. 'The market trend analysis and prospects of scaffolds for stem cells'. *Biomaterials Research*, 18, p 11. doi.org/10.1186/2055-7124-18-11.

Lundvall, B-Å, ed. 1992. *National Innovation Systems: Towards a Theory of Innovation and Interactive Learning*. London: Pinter.

Mallikarjun, Y. 2012. 'Stem cell therapy holds great potential', 25 January. Available at: www.thehindu.com/todays-paper/tp-national/stem-cell-therapy-holds-great-potential/article2829969.ece (accessed on 28 June 2015).

McMahon, DS, Singer, PA, Daar, AS & Thorsteinsdóttir, H. 2010. 'Regenerative medicine in Brazil: Small but innovative'. *Regenerative Medicine*, 5(6), pp 863–876.

Meissner-Roloff, M & Pepper, MS. 2013. 'Establishing a public umbilical cord blood stem cell bank for South Africa: an enquiry into public acceptability'. *Stem Cell Reviews and Reports*, 9(6), pp 752–763.

Mellet, J, Alessandrini, M, Steel, HC & Pepper, MS. 2015. 'Constituting a public umbilical cord blood bank in South Africa'. *Bone Marrow Transplantation*, 50, pp 615–616.

Ministry of Health and Family Welfare, 2015. Licensed Umbilical Cord Blood Banks in the Country. Available at: http://pib.nic.in/newsite/PrintRelease.aspx?relid=121409 (Accessed February 2018)

Narayan, P. 2010. 'New hope in stem cell therapy for blindness', 23 July. Available at: www.timesofindia.indiatimes.com/city/chennai/New-hope-in-stem-cell-therapy-for-blindness/articleshow/6202987.cms (accessed on 14 August 2013).

Nelson, R, ed. 1993. *National Innovation Systems: A Comparative Analysis*. New York/Oxford: Oxford University Press.

Nelson, RR & Rosenberg, N. 1993. 'Technical innovation and national systems', In Nelson, RR. ed. *National Systems of Innovation. A Comparative Analysis*. Oxford: Oxford University Press.

Niosi, J, Saviotti, P, Bellon, B & Crow, M. 1993. 'National systems of innovation: In search of a workable concept'. *Technology in Society*, 15(2), pp 207–227.

Pandeya, R. 2010. 'Drug controller queries stem cell therapy on spinal injuries'. *LiveMint*, 6 December.

Pandya, SK. 2008. 'Stem cell transplantation in India: Tall claims, questionable ethics'. *Indian Journal of Medical Ethics*, 5, pp 16–18.

Pepper, MS & Slabbert, MN. 2015. 'Human tissue legislation in South Africa: Focus on stem cell research and therapy'. *South African Journal of Bioethics and Law*, 8(2), pp 4–11.

Sachitanand, NN. 2009. 'Stem cell centre to rise in biology hub'. *Science*, 326(5958), p 1333.

Salter, B & Salter, C. 2010. 'Governing innovation in the biomedicine knowledge economy: Stem cell science in the USA'. *Science and Public Policy*, 37(2), pp 87–100.

Schwab, IR, Johnson, NT & Harkin, DG. 2006. 'Inherent risks associated with manufacture of bioengineered ocular surface tissue'. *Archives of Ophthalmology*, 124, pp 1734–1740.

Shankar, R. 2014. 'DCGI asks Chaitanya Hospital to stop treatments with stem cell products in name of clinical trial'. *Pharmabiz*, 17 December. Available at: www.pharmabiz.com/NewsDetails.aspx?aid=85692&sid=2 (accessed on 24 January 2017).

Singh, S. 2013. 'Stem cell industry: The battle within'. *Forbes India*, 12 February. Available at: www.forbesindia.com/article/real-issue/stem-cells-industry-the-battle-within/34697/1 (accessed on 24 January 2017).

Slabbert, MN, Pepper, MS & Mahomed, S. 2015. 'Stem cell tourism in South Africa: A legal update'. *South African Journal of Bioethics and Law*, 8(2), pp 41–45.

Subbu, R. 2015. 'Stempeutics gets Japanese process patent for stem cell drug'.

The Hindu, 10 June. Available at: www.thehindu.com/todays-paper/tp-business/stempeutics-gets-japanese-process-patent-for-stem-cell-drug/article7299758.ece (accessed on 11 September 2016).

The Hindu. 2006. 'Government to act against clinic using stem cell therapy'. Available at: www.hindu.com/2006/01/23/stories/2006012300720900.htm (accessed on 8 June 2010).

Thom, A & Low, M. 2016. 'Illegal experiment scandal rocks Free State Health'. *Spotlight*, 2 October. Available at www.spotlightnsp.co.za/2016/10/02/illegal-experiment-scandal-rocks-free-state-health/ (accessed 8 January 2017).

Tiwari, SS. 2013. 'The ethics and governance of stem cell clinical research in India'. PhD thesis. The University of Nottingham, United Kingdom.

Tiwari, SS & Desai, PN. 2011. 'Stem cell innovation system in India: Emerging scenario and future challenges'. *World Journal of Science, Technology and* Sustainable *Development*, 8(1), pp 1–23.

Tiwari, SS & Raman, S. 2014. 'Governing stem cell therapy in India: Regulatory vacuum or jurisdictional ambiguity?' *New Genetics and Society*, 33(4), pp 413–433.

Tiwari, SS, Raman, S & Martin, P. 2017. 'Regenerative medicine in India: Trends and challenges in innovation and regulation'. *Regenerative Medicine*, 12(7), pp 875–885.

Vemuganti, GK & Sangwan, VS. 2010. 'Affordability at the cutting edge: Stem cell therapy for ocular surface reconstruction'. *Regenerative Medicine,* 5(3), pp 337–340.

Wood, L, Haveman, J, Juritz, J, Waldmann, H, Hale, G & Jacobs, P. 2009. 'Immunohematopoietic stem cell transplantation in Cape Town: a ten-year outcome analysis in adults'. *Hematology/Oncology and Stem Cell Therapy*, 2(2), pp 320–332.

CONCLUDING REMARKS

Building the knowledge economy

Current strategies and developments in South Africa

ZAMANZIMA MAZIBUKO

IN THIS CHAPTER, THE KNOWLEDGE economy in South Africa and its links to emerging technologies for economic development are reviewed. A summary of selected government policies and strategies is presented to understand the advancement, or lack thereof, of science, technology and innovation (STI) by government. Furthermore, the arguments put forward in this book for challenges and opportunities for nanotechnology and biotechnology (including stem cell research) in South Africa are highlighted for possible input into national strategies.

STATUS OF THE KNOWLEDGE ECONOMY IN SOUTH AFRICA

Leading economies in the world have transformed over the past two decades into dynamic knowledge-based societies. Rapid technological changes, innovation and new knowledge have been the drivers of this growth and development in advanced economies. This has led to an inevitable and gradual shift from the traditional economy, which relies on labour and financial capital, to one reliant on the intellectual capital found in the knowledge economy. New knowledge is catalytic in translating economic activities into economic growth and increasing a country's global competitiveness. Ultimately, emerging technologies like nanotechnology and biotechnology have a role in the transformation into knowledge economies. However, without substantial investments in these emerging technologies and a highly skilled labour force, building a knowledge-based economy could prove to be problematic. Nanotechnology and biotechnology promise new and exciting possibilities but require extensive investment and synchronised, well-thought-out policies, as well as an active private sector for these promises to be realised.

Developing countries need to strategically position themselves in the knowledge economy, which relies heavily on a nation's research agenda and its alignment with its national system of innovation. The New Partnership for Africa's Development (NEPAD) African Innovation Outlook II survey of 2014 (NPCA, 2014) reports that African governments have gradually accepted the role of scienctific and technology-based innovation (STI) in the development of the continent and understand the connection between STI and sustainable economic growth, job creation and poverty reduction. However, it is not easy to predict exactly how a new technology will perform in the market and whether it will indeed achieve the results it promises. Despite this uncertainty, developing countries embrace emerging technologies because they present the possibility of leapfrogging to new industries while developed economies anticipate sustaining their competitive advantage.

Emerging technologies are expected to drive the industrial revolution in the 21st century, and in accordance with the South African National Development Plan (NDP), South Africa endeavours to be part of this pursuit of a forward-thinking global competitiveness. The NDP states that 'science and technology are key attributes of an equitable economic growth, because technological and scientific revolutions underpin economic advances, improvements in health systems, education and infrastructure' (NPC, 2011). The Ten-Year

Innovation Plan (2008) is one of a few plans drawn up by the South African government to advance STI and to help drive South Africa's transformation towards a knowledge-based economy.

The various available definitions of the knowledge economy are inconsistent and improvised according to specific needs. This makes it difficult to measure a country's knowledge economy. The OECD (1996) describes knowledge economies as those that are 'directly based on the production, distribution and use of knowledge and information'. This is associated with an increase in 'high-technology investments, high-technology industries, more highly-skilled labour and associated productivity gains'. Measuring the knowledge economy using the OECD definition involves the estimation of the value that knowledge- and technology-intensive industries and services (finance and insurance, telecommunications, business services) add in the country's GDP.

The World Bank (2012) uses a 'Knowledge Economy Index' as a composite indicator to measure the knowledge-based economy. This index bases the measure of a country's performance on four knowledge economy pillars, namely, economic incentive (which is an economic pillar) and three knowledge pillars comprising education and training; innovation and technological adoption; and communications technologies.

The innovation and education pillars are critical to the development of nanotechnology and biotechnology as emerging technologies. The variables in the innovation and technological adoption are:

- Royalty and license fees payments and receipts
- Scientific and technical journal articles
- Patent applications granted by the USPTO

The extent to which technologies are useful and functional – as is the case with nanotechnology and biotechnology (including human stem cell research and applications) – is dependent on coordination of policies including nationwide deliberations exploring issues of ethics. Feasibility, socio-ethical aptness and suitable intellectual property management are crucial in ensuring the success of a technology, especially one that is novel and undeveloped. The Centre for Science,

Technology and Innovation Indicators (CeSTII) is a statistical and policy research unit located within the Human Sciences Research Council (HSRC) that conducts national R&D and innovation surveys on behalf of the South African Department of Science and Technology. CeSTII produces national indicators from the survey results to provide inputs for policy-makers and a basis for international comparisons. According to their 2014/15 South African National Survey of Research and Experimental Development (2016), investment in R&D, specifically in biotechnology and nanotechnology, has been increasing over the years. The survey shows the 2014/15 investment in biotechnology being 5.4% and in nanotechnology 2.8% of gross domestic expenditure on research and development (GERD). Both biotechnology and nanotechnology grew 8.4% on average, faster than the compound average growth rate of GERD overall, which was also 8.4% in the same period.

The above statistics are encouraging. However, the education pillar of the Knowledge Economy Index (KEI) is disturbing. The South African Innovation Scorecard Framework (2016) reported that improvements in the education and training pillar were the lowest in 2014. The variables in the education and training pillar are:

- Secondary enrolment
- Tertiary enrolment
- Adult literacy rate

Looking at these variables with the knowledge of the poor state of education in South Africa, it is not surprising that education lags behind in the KEI. There needs to be more effort put into creating an enabling environment in which a science, technology, engineering and mathematics (STEM) education is attainable for the average citizen. This would facilitate the development of a generation that will reap the benefits of a knowledge economy.

AN OVERVIEW OF NATIONAL STRATEGIES AND POLICIES IN RELATION TO SCIENCE, TECHNOLOGY AND INNOVATION (STI)

The post-apartheid South African government swiftly began writing up various plans and/or policies aimed at reversing historical weaknesses and addressing the issues of development and inequality. This overhaul included science and technology. The White Paper on Science and Technology provided the roadmap for reviewing science and technology policy with the aim of establishing science and technology strategies and launching a successful National System of Innovation in South Africa. The use of technology and innovation to drive human development and empowerment and economic growth was emphasised.

More comprehensive reference to the national system of innovation and STI is first found in the national development plan (NDP), which was launched in 2012. Prior to the NDP, strategies such as the Reconstruction and Development Programme (RDP) and the Growth, Employment and Redistribution (GEAR) strategy mentioned the importance of science and technology in the development of several sectors and the need to support industrial innovation. However, these policies had no clear linkages to the national system of innovation (NSI) and STI. Rather, STI is referred to as a separate concept to research and development. The New Growth Plan (NGP) did, however, discuss the targeting of the creation of 100 000 new jobs in the knowledge-intensive sectors by 2020. This is one of a few areas that allude to the NSI and STI in this strategy.

The next part of this chapter looks at the location of NSI and STI in selected policies and the significant ideas in these policies that, if correctly implemented, could lead to the success of nanotechnology and biotechnology (among several emerging technologies) in the country's development plans.

New Growth Path (NGP)

As mentioned in the section above, the New Growth Path considers knowledge-intensive sectors as drivers for the creation of much-needed jobs. Biotechnology is one of the sectors identified as a strategic knowledge-intensive field that has potential for job creation. The need for large-scale job creation is emphasised and technological innovation is seen as the gateway to a possible solution to unemployment. The NGP targets '100 000 new jobs by 2020 in the knowledge-intensive sectors of ICT, higher education, healthcare, mining-related technologies, pharmaceuticals and biotechnology'. The strategy also identifies the following issues as important to strengthening STI and the NSI:

- Government and public-sector funding to ensure the creation of a considerable number of jobs and to keep existing and advanced industries alive
- A policy that allows for novel education and training systems (including institutions of learning within enterprises and state agencies) with increased R&D support
- A technology policy flexible enough to facilitate an environment conducive for research and growth in innovative technologies (including green technologies)
- Collaboration and coordination among relevant stakeholders

The NDP

The NDP recognises that a middle-income country such as South Africa requires innovation to have a global footprint. The NDP also points out that significant investment in research and development is necessary for South Africa to be a global competitor in the science and technology space. The document highlights the following as being crucial to the country's development and sustainable growth goals:

- A strong system of innovation that contributes to advances in technological innovation, the creation of new knowledge and consequently to transformation. This system of innovation must be coordinated with research institutions to advance South Africa's global competitiveness
- Government funding and support for research and new product development by existing industries and firms
- Partnerships and coordination between universities, science councils, departments, NGOs and the private sector including strategies and incentives to attract businesses to develop industry clusters designed to increase competitiveness and wealth

The importance of STI and the NSI in all sectors of the country is emphasised in the NDP, more than any other strategies before its launch. This means all government departments are implicated in ensuring the strengthening and success of STI as it has been identified as key to improving quality of life and improving economic competitiveness.

The national R&D strategy

The national R&D strategy identifies biotechnology as one of the scientific and technological R&D areas that require state resources. The strategy acknowledges the inadequate capacity in South Africa to respond to biotechnology as an emerging technology that is critical in the global economy. This lack manifests in both technological capacity and a failure to harness the social sciences to gain a holistic understanding of our system of innovation. As argued in the nanoethics chapter of this book, bringing on board the social sciences in the development of emerging technologies improves the rate of innovation in society. Additionally, South Africa has only limited protection of intellectual property related to new developments in biotechnology, which leaves the country vulnerable to exploitation.

The national R&D strategy aims to achieve the following:

- Reduce poverty and enhance quality of life using technological innovation
- Build key technology platforms that are new and knowledge intensive (National Biotechnology Strategy)
- Exploit the influence of existing resource-based industries and developing new knowledge-based industries from them

The Industrial Policy Action Plan (IPAP)

The Industrial Policy Action Plan (IPAP) is aligned to the National Development Plan (NDP) and is an integral part of the government's broad policy to manage key challenges of economic and industrial growth, including unemployment, inequality and poverty. The first IPAP iteration, informed by the National Industrial Policy Framework (NIPF), was approved by cabinet in 2007 and there has since been an annual update of this plan. The IPAP also supports the use of knowledge and technology-intensive solutions for the development and growth of the country. For example, DST's Advanced Manufacturing Technology Strategy (AMTS) aims to improve manufacturing in South Africa and advance its global competitiveness through innovation and advanced technologies that generate high-value products, contribute towards global supply chains, significantly increase export sales and have decent jobs with decent pay. Advanced manufacturing leverages on 'emerging physical and biological scientific capabilities' such as nanotechnology, biotechnology, chemistry and biology. The next industrial revolution (or the 4th Industrial Revolution) will aim to create with higher speed, at lower cost and greater precision at a molecular scale, which is the exact concept of nanotechnology.

The IPAP also recognises the importance of collaboration between the various departments in government, the private sector and academia (including science councils) to fully exploit STI for the country's development. The need for collaboration is a common theme in South Africa's policies and national plans. Other common themes that surface include:

- Emerging technologies as the key to development and transformation
- The importance of investment in technological innovation
- The creation of an enabling environment for SMEs to thrive
- A well-functioning national system of innovation is the key to South Africa's global competitiveness

These strategies are also synergistic with South Africa's Integrated Manufacturing Strategy (2003), which indicates that 'It is essential to develop a domestic capacity for science, technology development and advanced skill development'. The Manufacturing Strategy further states that, 'A new technology, innovation and research and development (R&D) strategy proposes a number of measures that government will take in this regard, including the provision of an enabling institutional environment for R&D, as well as developing South Africa's technological capacity in strategic areas such as biotechnology.'

Next to be discussed are the innovation policies specific to biotechnology, nanotechnology and building a knowledge economy.

National Biotechnology Strategy (NBS) and the Bio-economy Strategy

South Africa's National Biotechnology Strategy, which was launched in 2001, entailed the establishment of the biotechnology regional innovation centres (BRICs) together with a Biotechnology Advisory Committee (BAC) to assist with the success of biotechnology innovation. However, the strategy was reviewed after the BRICs were integrated with the Technology Innovation Agency (TIA) and not the BAC. In addition, the substantial investment into starting research companies yielded little to no results. In the wake of the failure of the NBS, the South African Bio-economy Strategy was launched in 2014. This strategy is headed by the DST with the TIA being responsible for facilitating the production and commercialisation of biotechnology products.

National Nanotechnology Strategy

The National Nanotechnology Strategy (NNS) endeavours to put South Africa on the global map in technological innovation. It discusses the convergence of nanotechnology, biotechnology and information technology which have the potential to solve global issues. South Africa is among the first countries to have an official strategy for nanotechnology, which was launched in 2005. The NNS recommends the establishment of nanotechnology characterisation centres, research and innovation networks, a capacity-building programme and a flagship project programme.

Ten-Year Innovation Plan

The Ten-Year Innovation Plan (TYIP) was published in 2008 to help transform South Africa into a knowledge-based economy where knowledge drives economic growth and human development. The plan aimed to contribute to the development of the NSI and to address the socio-economic issues of the country. The plan acknowledged that South Africa was still significantly behind in its attempts to develop into a knowledge-driven economy and to commercialise products from scientific research. TYIP focuses on developing new and innovative technologies and utilising multi-disciplinary approaches.

LESSONS AND RECOMMENDATIONS FROM BOOK CHAPTERS

The various chapters of this publication provide perspectives on the possible benefits and reservations of nanotechnology and biotechnology, including human stem cell research, and further reveal the gaps and problems that face policy-makers. Comparative studies have allowed for a broader analysis of these emerging technologies, looking at how South Africa can build its economy using these technologies while being cognisant of their potential risks. The recommendations provided by the authors could assist in identifying and filling knowledge gaps and in strengthening current policies.

In chapter 1: 'The advancement of nanotechnology: A sustainable development or an untenable vision?' the subject of technological determinism in nanotechnology is explored and the history of nanotechnology reviewed. The chapter looks at whether nanotechnology has a future or if it presents are fleeting promises. The author recommends research into the all-round sustainability of nanotechnology. This includes a scrupulous risk and life cycle analysis of nanotechnologies. Furthermore, the author recommends the involvement of society and governments in reducing the gap between the development of nanotechnologies and the nanoethics.

In chapter 2: 'Nanoscience, nanotechnology, nanomaterials and nanotoxicology in South Africa', the authors argue for the need to achieve a balance between the unfavourable and the beneficial effects of engineered nanomaterials (ENMs). Achieving this balance, fundamentally, would require rigorous regulatory systems and quality control, which are presently absent. Therefore, the expanding nanotechnology industry in South Africa must collaborate with risk management bodies. Several risk management gaps are identified and strategies are put forward as possible solutions for the current lack of risk assessment tools in testing the toxicity of nanomaterials:

- The need to understand the interactions between different types of ENMs (non-degradable or slowly degradable) and cells, in relation to their physicochemical properties is acknowledged. The authors also discuss the need for more extensive research on methods of administration, means of uptake, and the body's clearance mechanisms.

- The starting material and properties of nanomaterials must be identified, particularly for industrially produced ENMs where crude production processes have presented problems with impurities, for instance, which have led to subsequent variations in material properties. Knowing the preliminary material will eliminate the incorrect attribution of toxic effects to a certain properties of nanomaterials.
- Nano-related databases must be consolidated in order to create a definitive reference point.
- The nano-research community must formulate the best way to measure the toxicity of ENMs in studies and risk analyses.
- Occupational exposure limits for ENMs must be established and nanowaste management proposals must be strengthened to facilitate the rapid growth of nanotechnology.
- A precautionary approach should be introduced to safeguard both the environment and the health of researchers and workers handling nanomaterials while occupational exposure limits are established.

Chapter 3: 'Envisioning and engaging the societal implications of nanotechnology: Is it too early for Africa to do nanoethics?' probes the ethical issues surrounding the use of nanotechnology and the increasing technological knowledge gap between the developed world and poor African nations. The author makes the following recommendations:

- Africa needs to set the agenda and lead the debate on nanoethics as it is the continent with the most at stake.
- Nanoethics (through existing laws) should provide guidance for companies generating nanotechnology products to do so ethically.
- South Africa's nanotechnology policy should include promotion of the development of nanotechnology for the rest of Africa in order to allow the continent to benefit from South Africa's developed nanotechnology infrastructure and network. This could narrow the growing gap between the developed world, and African nations could also help other sub-Saharan countries to develop socially

inclusive nanotechnology governance aimed at the promotion of equality and equity.

Chapter 4: 'Disease of poverty: Nanomedicine research in South Africa' interrogates the relationship between private–public partnerships (PPPs) and nano-medicine development in South Africa. Globally, the involvement of PPPs in advancing medicines for diseases of poverty is growing exponentially and the chapter offers recommendations for how South Africa can participate in this sector:

- To improve South Africa's nanomedicine, the country must link its nanotechnology research efforts to the PPPs operating in poor and rural communities.
- Tight regulations in South Africa are slowing down the ability of PPPs to operate effectively, and will greatly hinder innovation in the country. Government needs to reconsider its regulations and policies to encourage research and innovation of PPPs.
- Government should provide additional funding for nanomedicine scholars to work with PPPs and should work with PPPs to develop policies that promote innovation and protect the public safety.

Chapter 5: 'Building a bio-economy in South Africa: Lessons from biotechnology innovation networks in taiwan' provides a comparison between Taiwan and South Africa's bio-economies. The author shows that Taiwan's foreign partnerships have played a positive role in developing the country's industries. The chapter recommends that:

- South Africa should implement more productive strategies to enhance research capabilities, help diversify their specialities, improve the research capacity of local academia, and support partnerships with global innovation networks.
- South Africa should build the local research capacity to develop and transform its nascent bio-economy.
- South–South partnerships and cooperation must be established.
- The South African government should ensure an effective coordination of research efforts via PPPs, and joint research and entrepreneurship programmes.
- South Africa should ensure that unnecessary regulations and

policies do not hinder innovation.

Chapter 6: 'What can South Africa learn from high technology patents in India: An analysis of biotechnology patents through USPTO' contrasts India's and South Africa's footholds in the biotechnology industry using an analysis of the biotechnology patents granted in USPTO. The chapter finds that there is a huge gap between South Africa and India mostly due to South Africa's strict regulations and high reliance on internal resources. The chapter recommends that:

- South Africa provides incentives for foreign firms to conduct their R&D operation in the country.
- South Africa develops a skilled workforce which could act as an incentive for foreign firms to start their R&D operation in the country.

Chapter 7: 'The proliferation of stem cell research and therapy in South Africa and India: A comparative study' reviews research into stem cell therapy in South Africa and India and recommends that:

- South Africa loosen its regulatory laws to stop the current stifling of product innovation caused by stringent laws.
- The South African government increases in public investment in research to create a productive innovation system that translates basic research into commercial success.
- South Africa needs to have a better scientific infrastructure including a robust regulatory framework to address unethical stem cell research.

CONCLUSION

South Africa has identified emerging technologies as key to development and transformation. There are still knowledge gaps that need to be filled, especially with regards to risk assessment of these emerging technologies. One way in which South Africa has embraced innovation and advanced technologies is the creation of national strategies. For example, the National Nanotechnology Strategy has encouraged the promotion of regional, national and international research partnerships in nanoscience; there are education and training initiatives including outreach programmes to create a skilled workforce and stimulate public discourse on nanotechnology.

Nanotechnology's potential to dramatically transform elementary technologies may create entirely new sectors. This is an exciting prospect but also presents policy challenges which are aggravated by the fast pace of technological innovation. It is clear that the regulation of nanomaterials has become imperative for policy-makers. As much as policies need to ensure the success of emerging technologies they must also protect society from the possible dangers presented by these technologies. Strong but flexible policies are crucial to building the bio-economy and to advancing nanotechnology. Policy-makers must be cognisant of the need to avoid the stifling of entrepreneurship and investments into emerging technologies by relaxing policies that are too stringent. It is a difficult balance but it must be achieved.

Fundamentally, South Africa and the rest of Africa are facing the challenge of catching up to the developed world. This gap applies not so much to core research and science aptitude as to the creation of an environment that enables innovation and allows for the commercialisation of emerging technologies. A robust system of innovation is an essential enabler of advancement in STI. Furthermore, there is great value to society in having its citizens informed and able to use technology to address the needs of that society. Policy frameworks should be vigorous yet nimble and they should be adjusted as the need arises, especially in the ever-changing world of technology.

REFERENCES

Centre for Science, Technology and Innovation Indicators (CeSTII). 2016. South African National Survey of Research and Experimental Development. Cape Town: HSRC.

DST. 2002. The National Research and Development Strategy. Pretoria: Department of Science and Technology.

DST. 2008. Innovation Plan: Toward the Knowledge Economy. Pretoria: Department of Science and Technology.

DTI. 2003. Integrated Manufacturing Strategy. Pretoria: Department of Trade and Industry.
DTI. 2011. Industrial Policy Action Plan 2. Pretoria: Department of Trade and Industry.
National Planning Commission (NPC). 2011. 'National Development Plan: Vision for 2030'. Pretoria: The Presidency.
National Advisory Council on Innovation. 2016. Pretoria: South African Innovation Scorecard Framework.
NEPAD Planning and Coordinating Agency (NPCA). 2014. African Innovation Outlook 2014. Pretoria: NPCA.
OECD. 1996. *Knowledge-Based Economy*. Paris: Organisation for Economic Cooperation and Development.
OECD. 1997. *National Innovation Systems*. Paris: Organisation for Economic Cooperation and Development.
Presidency. 2010. 'New Growth Path'. Available at: www.info.gov.za/view/DownloadFileAction? (accessed on 17 July 2017).
World Bank. 2012. 'Knowledge Economy Index'. Available at: data.worldbank.org/data-catalog/ (accessed on 19 July 2017).

Index

U

V

W

X

Y

Printed in the United States
By Bookmasters